Brain Power Enrichment
Level 3 Workbook
Student Version
Grades 8-10

A workbook for the development of
logical reasoning, critical thinking,
and problem solving skills

by
Reuven Rashkovsky and Karine Rashkovsky

AuthorHouse™
1663 Liberty Drive
Bloomington, IN 47403
www.authorhouse.com
Phone: 1 (800) 839-8640

© 2015 Reuven Rashkovsky and Karine Rashkovsky. All rights reserved.

No part of this book may be reproduced, stored in a retrieval system, or transmitted by any means without the written permission of the author.

Published by AuthorHouse   10/23/2015

ISBN: 978-1-5049-5814-1 (sc)
ISBN: 978-1-5049-5815-8 (e)

Print information available on the last page.

Any people depicted in stock imagery provided by Thinkstock are models, and such images are being used for illustrative purposes only.
Certain stock imagery © Thinkstock.

This book is printed on acid-free paper.

Because of the dynamic nature of the Internet, any web addresses or links contained in this book may have changed since publication and may no longer be valid. The views expressed in this work are solely those of the author and do not necessarily reflect the views of the publisher, and the publisher hereby disclaims any responsibility for them.

Table of Contents

Introduction 1

Lesson One 7

Lesson Two 13

Lesson Three 17

Lesson Four 21

Lesson Five 29

Lesson Six 35

Lesson Seven 41

Lesson Eight 47

Lesson Nine 51

Lesson Ten 57

Lesson Eleven 63

Lesson Twelve 69

Lesson Thirteen 75

Lesson Fourteen 79

Lesson Fifteen 85

Lesson Sixteen 89

Lesson Seventeen 85

Lesson Eighteen 101

Lesson Nineteen 107

Lesson Twenty 111

Lesson Twenty One 115

Lesson Twenty Two	119
Lesson Twenty Three	125
Lesson Twenty Four	129
Lesson Twenty Five	135

Introduction to Level 3 at Brain Power Enrichment Programs

Welcome to the Student Version of the Level Three course books for Brain Power Enrichment Programs. This exciting book is a natural continuation of the material included in Books 1 and 2 – the Student Versions of the Level Two course. This book accompanies a Level Three student through his/her problem solving program (or it may be used independently as a problem solving workbook) and it includes the materials utilized for a school year.

Problem Solving: What is it and why is it important?

Although problem solving may mean different things to many people in varying contexts, in the academic sense, problem solving can be defined as an attitude towards inquiry and an involvement in thinking operations of analysis, synthesis, and evaluation-- all of which are higher-level thinking skills.

The implications for improving one's problem solving skills are numerous. These include a more positive attitude toward math and science, improved thinking flexibility and creativity in all subject areas, as well as increased success on academic, university admissions, and professional program tests (many of which are designed with an emphasis on assessing higher-order thinking skills). Moreover, knowledge of a range of problem solving strategies coupled with experience in their application have benefits which transcend the classroom and enter the realm of professional, social, and intellectual accomplishment.

Course Objectives

Brain Power courses are guided and driven by the following principle: "giftedness" may be developed in most children with the combined team effort of students, parents and qualified teachers, through Brain Power Enrichment participation. By facilitating student problem solving ability and thereby enriching higher order cognition, Brain Power students acquire skills attributed to the academically gifted; namely, being able to grapple with intellectually sophisticated challenges, possessing the ability to integrate multiple ideas and facts, effectively and creatively finding solutions to dilemmas, and finding little difficulty with memory and attention to detail.

The objective of Brain Power Enrichment programs is to instill a love of learning through the development of a student's problem solving skills. The program aims to develop problem-solving abilities in students seeking to improve these skills, and/or provides challenge, stimulation and inspiration through engagement in problem solving for gifted students. Brain Power programs help children achieve academic success, obtain important learning tools, strengthen their thinking skills, and open doors for success in secondary, post-secondary and even graduate or professional programs.

Level 3

The course is structured to coincide with a student's regular school year (10 months). It consists of an average of 34 sessions/lessons; each lesson being two hours long and delivered once a week. Each lesson in this book includes material for a class session and homework, the latter of which is to be completed throughout the week between classes.

All Brain Power courses are based on a step-by-step approach, which enables students to understand problems of increasing complexity. Level 3 continues to equip students with various problem solving strategies and techniques, and supports the application of these skills to math, language arts, study habits, and the general learning process. In Level 3, students continue to sharpen their skills by applying four critical steps in problem solving. These steps and their corresponding explanations can be found below:

1. Understanding the problem
2. Defining a plan or strategy
3. Solving the problem
4. Checking the answer

The **understanding of a problem** continues to be the most critical component of the course at this level. This is the least defined and the least tangible step, but failure in understanding is fatal for problem solving. The recommendations for this step include:

Level Three

- Reading the problem several times
- Observing visually
- Listening carefully to teacher's and/or other student's explanations
- Deciding what to look for: defining important information, eliminating unnecessary information
- Simplifying the problem if appropriate (there are several techniques for problem simplification which will be covered)
- Modeling the problem
- Discussing the problem with peers, with parents, and with teachers

This process must be carefully monitored by the teacher every lesson throughout the school year. Reminders and reinforcement through practice will ensure students pay due diligence to this step in problem solving.

The second step in problem solving is **developing a plan or strategy**. This course continues to reinforce strategies learned throughout Levels One and Two of the Brain Power Problem Solving program.

The third step involves actually **solving the problem** and articulating the answer.

The last step, which is usually neglected by many students, is to **check** if the answer is appropriate for the problem. This step is particularly important while solving algebra problems.

These four important steps are applicable to analytical and creative problem solving. The course includes analytical problem solving requiring math, logic, and language arts. In addition, the course includes some topics from science, creative problem solving, as well as memory development.

Typically, a Brain Power student's critical thinking abilities will improve in the first three to four months of diligent participation in his/her Brain Power course. This can be measured by the student's improvement in her/his academic performance, as well as increased self-confidence in problem solving methodology and practice.

(*Note: It is strongly recommended that calculators are not to be used by students at this level).

The Roots of Brain Power

The Brain Power Enrichment courses were first conceived of and developed in 1974, within the European educational milieu of this time. The program has been updated yearly since the beginning in the 1980's to reflect and respond to the urgent necessity of North American educational supplementation. The courses have been continuously developed in response to many parents' concerns with the public, and most private educational systems, which do not emphasize or provide enough (or, sadly, sometimes any) development of thinking and problem solving skills. This course has been successfully supplementing the regular school curriculum in order to maximize children's abilities in problem solving for over two decades. In doing so, the programs have fostered academic growth, facilitated student acceptance to gifted, prestigious and professional programs, and generally increased hundreds of students' confidence in their critical thinking skills.

More information about Brain Power Enrichment Programs may be found on the web-site: www.besmarter.ca or www.mybrainpower.ca

Students

This Level Three course is intended for children aged 13 – 16 (students in grades seven to ten).
The most effective delivery of this course is by an experienced teacher who is trained in problem solving, and the material that should be used with small classes (8 – 12 children per class) of relatively homogenous abilities.

Students should be encouraged to independently attempt various methods for solving problems as opposed to attempts to rationalize their thinking based on known answers. Finally, it is important to continuously demonstrate to students that there may be more than one way to solve a particular problem, and, more importantly, to emphasize that the most "elegant" solution is the easiest and simplest one.

Level Three

Level Three

Level Three

LESSON 1

Classwork:

1. Continue the following patterns:

   a. $\frac{1}{2}$, 1, 2, 4, 8, __16__, __32__, __64__, __128__

   b. 3, 1, 5, 3, 7, 5, 9, 7, __11__, __9__, __13__, __11__

   c. 1, 3, 4, 7, 11, 18, __29__, __47__, __76__, __123__

   d. 1, 4, 9, 16, 25, __36__, __49__, __64__, __81__

   e. 2, 41, 5, 38, 8, 35, 11, __32__, __14__, __29__, __17__

2. Calculate/Solve:

   a. $\frac{3}{4} - 5\frac{1}{3} - (-\frac{5}{6}) =$

   b. $(-\frac{5}{7})^2 - (-\frac{8}{21}) =$

   c. $\{ -6[25 - 6^2 - 3(9^2 - 8^2) + 15] - 54 \div (-9) \} \cdot (-3) + (-25) =$

   d. $5X = 127$ $X - 5 = 127$

   $5 - X = 127$

   e. $5X + 87 = 12$ $5(2 - 6X) = 32$

   $2X - 3(4X - 7) = 3X$

3. Read the story below and then evaluate the statements about this story. Circle one of the letters, T, F, or X. T- stands for True, F – stands for False, X – stands for "I am not sure".

   "A student ran into a room but it was too late; an angry Professor had already stuck a sharp knife into the heart of a beautiful girl. Two minutes later, a police officer entered the room and took the evidence away. The man was arrested the next day."

   Statements:

   a. The man who killed the beautiful girl was a professor. (T) F X

   b. The Professor was angry with the girl so he killed her with a sharp knife. (T) F X

   c. A student ran into the room to prevent a crime. T F (X)

   d. It took the police only 2 minutes to arrive since somebody placed a phone call. (T) F X

   e. The story involves four living men and one dead girl. (T) F X

   f. The man who was arrested the next day was a Professor. (T) F X

   g. The man was arrested based on the evidence taken by the police officer. (T) F X

Level Three

4. **Four partners, Fred, Frank, Freud, and Fram,** are going to the meeting room to discuss the business of their company. They occupy positions of CEO, COO, VP of Sales, and CFO but not necessarily in that order. Noticing that Freud is not here as he is always late for meetings, Fram mentioned to the COO that nobody in Freud's family was a psychologist. Fred gave the VP of Sales and the CFO the latest economic forecast. The CFO offered Frank to give a presentation to his department next week.
What position in this company does each person occupy?

5. Write values for each sequence below:

   a. Term # times (-3), plus 5:

   Term #: 1st 2nd 3rd 4th 5th

   Term: __ __ __ __ __

   b. Term # in power 3, minus 35:

   Term #: 1st 2nd 3rd 4th 5th

   Term: __ __ __ __ __

   c. Term # plus next term #:

   Term #: 1st 2nd 3rd 4th 5th

   Term: __ __ __ __ __

   d. 11 take away the term #:

   Term #: 1st 2nd 3rd 4th 5th

   Term: __ __ __ __ __

6. **Nicholas is working part-time** in a restaurant. Today he is washing dishes by hand. There were 18 dirty plates in the sink when he started working. Nicholas works very fast - he cleans 5 plates every minute. However, the waiter adds two more dirty plates into the sink every minute. How many plates in all will Nicholas wash in 10 minutes?

7. **Rachel, Elliott, and Gabriel** collected 100 ancient coins altogether. The number of coins collected by Rachel and Gabriel together is 80. The number of coins collected by Elliott and Gabriel together is 40. How many coins did each of them collect?

8. **Rachel, Elliott, and Gabriel** skipped last week's quiz in math. Each of them gave a "story" to their teacher on why they could not participate in the quiz. When challenged by the teacher, Rachel said that Elliott lied, Elliott said that Gabriel lied, and Gabriel said that both Rachel and Elliott lied. The teacher knows that only one of these three students is telling the truth. Which of these three students is the truth teller?

9. **Anna's age is 2 years more** than twice Boris's age. Will's age is 3 years less than four times Boris's age. The sum of Anna's age and Will's age is 71. How old will Boris be in 5 years?

Level Three

10. **Four students, Rachel, Elliott,** Gabriel, and Sam, bought 105 batteries in all. Each of them bought a different number of batteries. If Rachel would buy only half ($\frac{1}{2}$) of the batteries she purchased, and if Elliott would buy $\frac{1}{3}$ of the batteries he purchased, and if Gabriel would buy 2 less batteries, and if Sam would buy 2 more batteries, then each of them would buy the same number of batteries. How many batteries did each of them buy?

11. **Eden is calling her mom** on her cell phone. She is telling her mom that she is standing in line for tickets for an amazing Rock concert. When her mom asked Eden how long she will be standing in line, Eden said that the line is long, but she is just in the middle of the line. Eden said that she is 50th from the front of the line, and she is 50th from the last person in the line. How much money will be collected for the concert if each ticket is priced at $25? (Assume that no more people will join the line).

Level Three

Fun Home Assignment:

1. **In the old days there were only balance scales** without weights. There are three coins and it is known that one of the coins is fake and is slightly lighter than the other two coins.
   A. How can you figure out which of these three coins is fake using the scales only once?
   B. What is the minimum number of times needed to use these scales to find one fake coin (lighter than the others) out of four coins?

2. **Ilana visited the Zoo last week.** She really liked the monkeys in the Zoo and she is sharing her impressions with her friends. She remembers seeing gorillas, chimpanzees, and orangutans, 17 monkeys in all. Ilana said that the number of gorillas she saw is four times less than the number of chimpanzees. She does not remember exactly how many orangutans there were, but she is sure that the number of orangutans there are is a prime number. How many gorillas, chimpanzees, and orangutans did Ilana see at the Zoo?

3. **There are four natural numbers that** the sum of these four numbers equals exactly the product of the same four numbers. What are these numbers?

4. Solve:

   a. $0.25X - 1.3 = (-0.775)$

   b. $3\dfrac{1}{4} - \dfrac{4}{5}X = 3\dfrac{1}{100}$

   c. $15 - 7(3 - 2X) = 10.8$

   d. $8.5(5X + 12.3) - 3.4(7 - 3.1X) = -9.418$

   e. $15 - 8(5X - 4) = 7.5X - 105$

5. **Dennis likes to pretend to be a magician.** Today he brought a new trick to his school. He wrote numbers from 1 to 9 on the whiteboard and he asked his classmates to choose one of these numbers. Julia chose 6. Then Dennis wrote 18 under the number 6 and he asked Julia to multiply 18 by 37. To her surprise, Julia's product was 666! When Henry said that he likes the number 4, Dennis wrote 12 underneath the number 4 and then multiplied it by 37. This time the answer was 444! Then Dennis announced that the product of any one digit number with 3, multiplied by 37, will produce a three digit number formed by the same digits as the one digit number that was chosen initially. How can you explain this "magic"?

6. **Fill up the empty squares with the appropriate digits:**

```
        7 □
     X □ □
     -------
       □ □
     □ □ □
     -------
     □ 2 □
```

Level Three

7. **On his birthday, Daniel decided to buy muffins** for the whole class. But, in addition, he bought chocolate candies for his best friends, 4 per friend. Daniel bought 44 items in all. How many students are in Daniel's class if one seventh of his classmates are his best friends?

8. **Jeremy and Bernie together** earned the same amount as Barbara and Aliza together. Bernie earned $252 and Aliza earned $307. Who earned more, Jeremy or Barbara and by how much?

9. **Use any four arithmetical operations** (+, -, x, ÷) and brackets as many times as needed to make the following equalities correct:

    2   2   2 = 2

    2   2   2 = 3

    2   2   2 = 11

    2   2   2   2 = 5

    1   1   1   1   1 = 100

10. **Ariel is delivering newspapers** in his neighborhood every Saturday morning. Last Saturday, his clients complained that page 9 from the paper was missing. What other three pages from that newspaper were missing that Saturday if there were 32 pages in the whole Saturday paper?

11. **Five friends, Avital, Laura, Sam, Victor, and Jim,** are discussing their birthday plans for this year. Please find out from the following clues the age of each co-worker and the month in which each of them will celebrate their birthday:

    a. Avital's birthday is not in April

    b. Sam is not the youngest and is not the oldest. He is one year older than Victor and his birthday is later in the year than Victor and Laura's birthdays

    c. Laura is younger than her friend who celebrates her/his birthday in May

    d. Avital is two years older than her friend who's birthday is in January

    e. Jim is 2 years older than Avital

    f. Jim's birthday is three months after Laura's

    g. The 27 year old is celebrating his/her birthday in March

	B-Days					Ages				
	Jan.	Feb.	Mar.	Apr.	May	25	26	27	28	29
Avital										
Laura										
Sam										
Victor										
Jim										

Level Three

Level Three

LESSON 2

Classwork:

1. **Solve the equations:**

   a. $5(7 - 3X) - 9(2 - 4X) = 50 + 10X$

   b. $\dfrac{3X - 5}{3 - 7X} = \dfrac{-23}{45}$

   c. $\dfrac{3.2 - 0.5X}{0.3} = \dfrac{5X + 8.95}{3.3}$

2. **Three students, Ron, Tom, and Ben** are members of the Problem Solving club. Their last names are Brown, Black, and White. Ron's last name is not White. Tom is in grade 8 this year and his father is a computer programmer. White is in grade 9. The father of the student Brown is a pilot. Figure out each student's full name.

3. **There are three shelves on a wall.** The number of books on the bottom shelf is one more than three times the number of books on the top shelf. The number of books on the middle shelf is six less than twice the number of books on the bottom shelf. How many books are on each shelf if there are 57 books in all on all three shelves?

4. **Mr. Shuster is a physician.** Today he examined four patients: Mr. Green, Mr. Roberts, Mrs. Sweet, and Mrs. Bridge. Dr. Shuster was amazed to find out that their ages are consecutive odd numbers; Mr. Green is youngest, then Mr. Roberts, then Mrs. Sweet. Mrs. Bridge is the oldest amongst them. The age difference between three times Mr. Green's age and twice Mrs. Bridge's age is 25 years. What is the age of each patient?

5. Find values of angles X, Y, Z

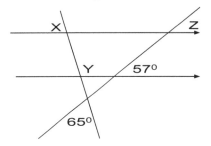

6. **Yuval bought three pens** and two highlighters for $2.20. Liran bought two pens and three highlighters for $2.30. What is the price of one pen and one highlighter?

7. **How would you place six chairs** in a rectangular room, having two chairs at each wall?

8. **There are 35 students in grade 8.** If there are 7 more girls than boys, how many girls are there?

9. **There are 35 students in grade 9.** How many girls are there if the ratio of boys to girls is 3 to 4?

Level Three

10. In △ABC, ∠A is 14° less than ∠B. ∠BCD = 132°.
Find the values of all three angles in this triangle.

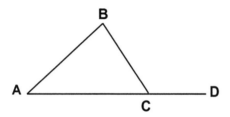

Level Three

Fun Home Assignment:

1. **A pack of 8 cans of yogurt** costs the same as 1 container of sour cream. One container of sour cream and one can of yogurt cost $4.23. I bought 2 containers of sour cream. How much did I pay?

2. Solve the equations:

   a. $12(32 - 7X) - 8(5 - 11X) = 110X + 26$

   b. $1.7X - 4.5(2.8X - 7.6) - 6.1X = 52.9$

   c. $\dfrac{4.7 - 2.7X}{0.5} = \dfrac{8X - 2}{-2}$

   d. $\dfrac{2.3X - 7.5}{3.6 - 5.7X} = \dfrac{35}{213}$

3. **One day mom gave Jennifer money** to buy a lunch at the school's cafeteria. When Jennifer returned from school, her mom asked if Jennifer has any change from the money mom gave her in the morning. Jennifer answered:
   - With half of the money you gave me this morning I bought a sandwich
   - One fifth (1/5) of the money I spent on apple juice
   - Three tenths (3/10) of the money I spent on chewing gum.

   How much change did Jennifer give her mom that day?

4. **A group of 40 soldiers** need to cross a wide river. The bridge over the river was broken and the water was extremely cold. There was a little row boat on the shore, but this boat could carry only one soldier at a time. The officer noticed two boys playing soccer not too far away. He called these two boys to help him solve the problem. With the boys help, the officer found out that the boat can carry two boys, but not a boy and a soldier. After some thinking, the officer devised a plan to transfer his soldiers over the river using the help from these two boys and this small boat. What was the officer's plan? How many trips over the river did the boat make in all?

5. **Max, Sam, and Yoad are collecting** money for the school trip. By now, Max has five dollars more than three times Sam's amount of money. Yoad has $19 less than Sam and Max together. How much money do each of them have, so far if they have $127 altogether?

6. In $\triangle ABC$, $\angle A$ relates to $\angle B$ and to $\angle C$ as $3 : 7 : 5$. Find values of all three angles of this triangle.

7. **How would you place 16 chairs** in a rectangular room, having 5 chairs at each wall?

8. **Jordan works in a furniture factory.** Today he is making a delivery to the furniture store. The number of chairs he delivered today is 7 less than nine times the number of tables. The number of cabinets is 2.5 of the number of tables. How many of each kind of items did Jordan deliver today if the total number of items was 218?

Level Three

9. **The sum of twice of the smallest** of four consecutive even numbers and three times the third number is 172. What is the fourth number of these consecutive even numbers?

10. **On the first day of school,** Emily and Amy went shopping for school supply items. That day Emily bought two binders and one notepad for $8.13 but, on the same day, Amy bought three binders and two notepads for $12.78. Can you find out how much one binder costs without using equations?

11. **Find values of angles X, Y, Z:**

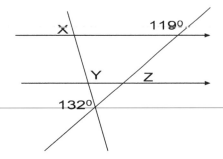

12. **How would you place 10 chairs** in a rectangular room having the same number of chairs at each wall when all chairs touching walls?

13. Four students from Mrs. Cowan class, Black, Brown, White, and Jackson, won regional competitions in math, poetry, physics, and arts. From the clues below, please determine each winners' full name and the subject in which the student won first place:

- Eden and the winner in poetry sit beside each other in the class

- Black did not win in physics nor in math

- Neither Elina nor Brown participated in the poetry competition

- Brown and the winner in arts were invited to Elina's birthday party last year

- Black's first name is not Eden nor Liran

- Liran and Jackson did not participate in the arts competition

- Eden did not win the math competition

- Jackson did not win the math nor the poetry competition

- Sophia slept well the night before her competition

14. **Three mothers and three daughters** each bought a $550 evening dress for a wedding party in their family. What is the smallest possible amount that they paid for the dresses?

Level Three

LESSON 3

Classwork:

1. Solve the equations:

   a. $\frac{1}{5}(\frac{1}{2}X+4) = \frac{1}{3}(\frac{1}{4}X+3)$

   b. $\frac{3X}{X-8} = 11$

   c. $\frac{3X-1}{5X+4} = 4$

   d. $\frac{2X-5}{3X+7} = \frac{11}{2}$

2. **In 2012, Ron was 18 years old** and Tom was 24. In what year was Tom three times older than Ron?

3. **Aliza's grandmother is one year less** than eight times Aliza's age now. In five years, Aliza will be five times younger than her grandmother then.
   What is the sum of their ages now?

4. **Jarod is three times younger than Alex.** In seven years, Alex will be four years older than twice Jarod's age then. Alex's mom is twenty five years older than Alex. How old is Alex's mom now?

5. **There are five parking spaces reserved** in front of the office building for five executives of the company, CEO – Mr. McKay, COO – Mrs. Fox, VP Marketing – Mrs. Davis, VP HR – Mr. Peters, and VP PR Mrs. Strong. Their cars are in front of the building now: gray Toyota, gray Honda, black Toyota, black Honda, and red Honda. Please find out who owns which car and in which parking space they park their cars in.

   . Mrs. Davis does not drive a Honda
   . The colour of Mrs. Fox's car and Mr. McKay's car is the same
   . None of the Honda cars are parked beside each other
   . Mr. Peters' car is parked between Mrs. Davis' car and a grey car
   . The car in the first parking spot is not red and not grey
   . No two cars with the same colour are parked beside each other
   . Mrs. Fox's car is not parked in the fifth parking spot

	Gray Toyota	Gray Honda	Black Toyota	Black Honda	Red Honda	1	2	3	4	5
McKay										
Fox										
Davis										
Peters										
Strong										
1					
2					
3					
4					
5					

Level Three

6. **Prove that the side against the 30° angle** in a right triangle equals half of the length of the hypotenuse in this triangle.

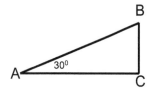

7. **Construct a perpendicular line through the mid-point of segment AB:**

A _____ B

8. **Four students, Alex, Adam, Anna, and Aaron, contributed $32 in all** to buy a present for their friend Boris who won a Problem Solving competition. If Alex would contribute $3 more, and Adam would contribute $3 less, and Anna would contribute three times less, and Aaron would contribute three times more, then all of them would contribute the same amount. How much money did each of them contribute?

9. **Danielle and Rachel have the same amount of money.** How much money should Danielle give to Rachel so that Rachel will have $15 more than Danielle?

10. **Linda is twice older than her daughter Beata** and three times older than her son Joseph. The sum of their ages now is 77 years. What was Joseph's age when Linda was four times older than Beata?

11. **Do you remember that Kurdu Murdu island is inhabited by two types of aliens: liars and truth tellers?** One day 175 Kurdu Murdians attended a reception at the island's chief palace. To their surprise, they found out that at least one among them was a truth teller. As well, given any two Kurdu Murdians who attended this reception, at least one of them was a liar.
How many truth tellers and how many liars attended this reception?

Level Three

Fun Home Assignment:

1. **Solve the equations:**

   a. $\dfrac{15}{2X-7} = \dfrac{4}{5-X}$

   b. $1.5(3.2X-6) - 4.1(2-5.4X) = 9 - 2.3(6.3-1.5X)$

   c. $\dfrac{7X-3}{8} = \dfrac{2X+5}{2}$

   d. $\dfrac{3.1-5.2X}{6.4X+8.7} = \dfrac{2.9}{7.3}$

2. **In the diagram below** lines a and b are parallel and the measures of two angles are known (132^0 and 123^0). What is the measurement of angle X?

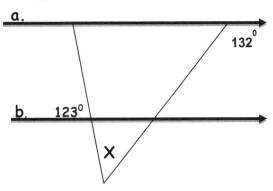

3. **Farmer Donald hired two farmhands**, John and Jack, for the summer season. He offered them two pay options: either $12 per each day worked or $1 on their first day, $2 on their second day, $3 on their third day, etc..., adding one more dollar for each day worked. John did not want to think long and he agreed for $12/day. Jack, after a few minutes of thinking, chose the second pay method.

   a. Which of these two farmhands will earn more money and by how much if they worked 45 days in all that summer for farmer Donald?

   b. On what day from the beginning of this season will John's total earnings be the same as Jack's?

4. **Prove that:** a. diagonals in rectangle are equal and b. point M, intersection of diagonals, is splitting each diagonal into two equal segments.

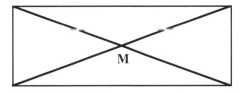

5. **Gabriel opened his piggybank** as he needs to buy a gift for his girlfriend Rebecca. There were dollar coins, quarters, dimes, and nickels. Gabriel counted the total amount of $41. How many coins in total were in Gabriel's piggybank if there were twice more dimes than nickels, eight more quarters than dimes, and the number of nickels was two more than twice the number of dollar coins?

6. **Four schools took part in a Problem Solving** competition. Each school's delegation included one teacher and three students: two girls and one boy. It was agreed that the captain in each team must be a girl. It happened to be that the names of each team member began with different letters: L, J, M, K. Please figure out each team member's names based on the clues below:

Level Three

a. A girl's name in Jane's team starts with the same letter as the name of the boy in Mr. Moore's team
b. Kent's teacher's name starts with the same letter as the name of Mark's captain.

Teachers	Captains				Boys				Girls			
	Leah	Janne	Mindy	Kathy	Leo	Jack	Mark	Kent	Lina	Josie	Mina	Karol
Levy												
Jones												
Moore												
Kramer												
Girls — Lina												
Josie												
Mina												
Karol												
Boys — Leo												
Jack												
Mark												
Kent												

You can use the extra clue below if the problem is too difficult for you:

(Please, try to solve this problem without the clue "c" first)

c. Jane's teacher's name does not start with the letter "L".

7. **Three years ago Sam was four times** younger than his sister Emily. Next year Sam will be twice younger than Emily. How old is Sam now?

8. **Amy is five years older than Karolina.** Fourteen years ago she was twice as old as Karolina then. What was the sum of their ages ten years ago?

9. **Prove that parallelogram ABCD** is a rectangle if its diagonals are equal in length.

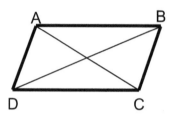

10. **Given right triangle ABC.** This triangle is isosceles; sides AC and BC equal 18 cm in length. Rectangle CDEF shares angle C with the triangle ABC and vertices d, e, and f are on the sides of the triangle. (I.E. rectangle **CDEF is inscribed into triangle ABC**). Find the perimeter of rectangle CDEF.

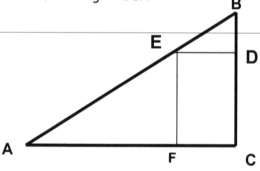

11. **The Ages of Mark and Julia** are consecutive even numbers. Fourteen years ago Mark was twice older than Julia then. What will the sum of their ages be next year?

Level Three

LESSON 4

Classwork:

Base Egyptian Numerals:

1 = |

10 = ∩

100 = ℮

1000 = 𝍌

10000 = ⟨

100000 = 𝍏

1000000 = 𝍐

1. **Translate from Egyptian Numerals** to Hindu-Arabic notation:

   a. 𝍏𝍏 ℮℮ 𝍌 ∩|||| =

   b. 𝍐𝍐 𝍏 ℮℮ 𝍌 ∩|| =

   c. 𝍐 ⟨⟨⟨ 𝍏 ℮℮ 𝍌𝍌𝍌 ∩ |||||||| =

   d. 𝍐 ⟨⟨⟨⟨ ℮℮℮℮ 𝍌𝍌𝍌 ∩ |||||| =

2. **Translate from Hindu-Arabic** notation to Egyptian numerals:

   a. 78 =

   b. 685 =

   c. 549723 =

   d. 7372046 =

3. **Reverse the pyramid** by moving only three circles:

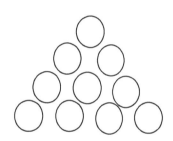

4. **Add/Subtract in Egyptian Numerals:**

   a. 𝍐 ⟨⟨ 𝍌𝍌𝍌𝍌 ∩ 𝍏 |||
      + 𝍏 ⟨⟨⟨⟨ ℮℮℮ 𝍌𝍌𝍌 ∩ ||||||
      ─────────

   b. 𝍐 ⟨⟨ 𝍌𝍌𝍌𝍌 ∩ 𝍏 |||
      − 𝍏 ⟨⟨⟨⟨ ℮℮℮ 𝍌𝍌𝍌 ∩ ||||||
      ─────────

Level Three

5. **Farmer John noticed** that his neighbor's sheep are coming on his territory to graze grass. So, to stop this without quarrelling with his neighbor, John decided to put a long fence with posts 6 meters apart. However, the lumber shop had 5 posts less in inventory than the number of posts John needed. Farmer John bought whatever number of posts that was available at the lumber shop and he built a fence of the same length but he put the posts 8 meters apart. How long is farmer John's fence? How many posts did John buy?

6. **Sebastian took out $1560** from an ABM (Automatic Banking Machine). When counting the money, Sebastian noticed that the whole amount was given in $50 bills, $20 bills, and $10 bills. The number of $20 bills was two more than twice the number of $50 bills. The number of $10 bills was two more than three times the sum of $50 bills and $20 bills. How many bills in all did Sebastian get from the ATM?

7. **The number of quarters** in a piggy bank is 6 more than twice the number of dimes. The number of nickels there are, is one third of the number of dimes. How many of each type of coin are there, if the total amount in the piggy bank is $18.15?

8. **Three friends, Willy, George, and Ben**, are riding motorcycles on a highway. One friend is riding a white motorcycle, the other friend is riding a grey motorcycle, and the third friend is riding a green motorcycle.
"Hey guys, did you notice that none of the first letters of our names are the same as the first letters of the colour of the motorcycle we are riding on?" exclaimed the rider of the green motorcycle. "Oh, yes!" replied George and Will. Who is riding which motorcycle?

9. **Fill in all the missing digits:**

				☐	☐	☐
×						
				6	☐	☐
			2	☐	1	☐
		☐	☐	☐	3	☐
	☐	☐	3	☐	☐	☐
3	☐	1	6	☐	5	

Level Three

10. **There were 150 participants** at the Linguistic convention. Twenty participants did not speak English nor French, 80 spoke English, and 87 spoke French. How many participants spoke both, English and French?

11. **The total amount of money Julia had** in her heavy purse was $15.25. Julia had loonies, quarters, and nickels in the purse. The number of nickels was one more than twice the number of loonies, and the number of quarters was twice the number of nickels. How many coins in all were in Julia's purse?

12. **One angle in a parallelogram** is 40^0 larger than another. Calculate all angles of this parallelogram.

Level Three

Fun Home Assignment:

1. **Translate from Egyptian Numerals** to Hindu-Arabic notation:

   a. [Egyptian numerals]

   b. [Egyptian numerals]

   c. [Egyptian numerals]

   d. [Egyptian numerals]

   e. [Egyptian numerals]

2. **Last week, a store owner received a supply** of cans of pickled cucumbers in 7 large boxes and 3 small boxes. An invoice indicated that there were 108 cans altogether in these boxes. This week the store owner received 5 large boxes and 4 small boxes of pickled cucumbers. This week the invoice states that there are 92 cans of pickled cucumbers altogether. The store owner does not want to open these boxes but he is keen to know how many cans are in each large box. Despite his lack of knowledge in algebra, the store owner managed to figure this out. Can you figure out how many cans are in each large box?

3. Add/Subtract in Egyptian numerals:

   a. [Egyptian numerals addition]

   b. [Egyptian numerals addition]

   c. [Egyptian numerals subtraction]

   d. [Egyptian numerals subtraction]

   e. [Egyptian numerals subtraction]

Level Three

4. A senior officer of the ancient Egyptian chariot platoon writes to his general: "O great general Phohoon! I am puzzled and therefore I need your help. Yesterday, my soldiers captured an enemy's village and they found ⅠⅠ barrels of wine there. One barrel contained ⅠⅠⅠ times more wine than another. My soldiers added some wine from our own supplies to these barrels: ∩∩∩∩∩ litres from one of our barrels and ℮∩∩∩∩ⅠⅠⅠⅠ litres from another of our barrel. Now both captured barrels contain the same volume of wine – ℮∩∩∩∩∩ⅠⅠⅠⅠⅠ litres each. According to your order, we must give ⊂ of the original content of the large barrel of wine we've captured to the chief Priest and another ⊂ to the Pharaoh. Please ask your scribes to calculate how much wine we should deliver to the Priest and how much to the Pharaoh." **Imagine you are an Egyptian scribe – how would you calculate these amounts?**

5. Three grade 7 students, Emily, Darya, and Elina, saved money last year to donate to the Hospital for Sick Kids. Emily collected quarters, Darya collected dimes, and Elina collected nickels. After counting their money, Emily was pleased that she collected $1.50 more than twice the amount collected by Darya. Elina was also proud because she collected $7.60 more than Darya. How many coins did each girl save last year if the total amount donated by the three of them was $173.10?

6. A local theater sold all tickets for the matinee show. The senior's tickets were sold for $8.50 per ticket, the children's tickets were sold for $5.50 per ticket, and regular tickets for adults were sold for $14 per ticket. The total amount for all the tickets sold was $5920.50. This was the record revenue for the matinee show this season. How many senior's tickets were sold and how many children's tickets were sold for this show if the number of senior's tickets was 34 less than twice the number of children's tickets, and the number of regular tickets was 23 more than four times the number of children's tickets?

7. Fill in the missing numbers:

		X	Y	Z
	×			
		Y	X	Z
			X	
		Y		

8. Given triangle ABC. M is a mid-point of the side AB, N is a mid-point of the side BC, and P is a mid-point of the side AC. A line is drawn through point M, perpendicular to side AB; through point N is a line drawn perpendicular to side BC, and through point P is a line drawn perpendicular to side AC. **It is a known**

Level Three

fact that all three perpendicular lines drawn through the mid-points of the three sides of any triangle intersect in one point, O. Please prove that point O is the centre of the circle that inscribes the given triangle (i.e. all three vertices of the triangle are on the circumference of this circle).

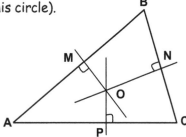

9. **Sixty one (61) hamburgers** were prepared for a trip to the ocean for 10 people. Each man took 7 hamburgers and each woman took 4 hamburgers. How many women went on this trip? Can you solve this problem without equations (without algebra)?

10. **The box taken out of the parking meter** was full of coins: toonies ($2), loonies ($1), quarters, dimes, and nickels. The total amount of money in this box was $67.50. How many coins in all were in this box if the number of loonies was 2 more than twice the number of toonies; the number of quarters was 4 more than 4 times the number of loonies; the number of dimes was 3 more than 5 times the number of loonies, and the number of nickels was 2 more than the sum of all toonies, loonies and quarters?

11. Five friends, **Ben, Ron, Bob, Leo, and Ken** live on the same street in houses numbered 1, 3, 5, 7, and 9. Each of these friends live in a different house and each of them have a different pet. Please figure out who lives in which house and what pet each friend has:
   a. Ken's pet is not a cat
   b. There are two houses between Ken's and Leo's houses
   c. Ben's house number is higher than the house number where a goat is the pet
   d. Bob lives between Ken's and Leo's houses
   e. The cat owner lives in the house with the street number lower than the house where Ron lives, but higher than the street number where the owner of a bird lives
   f. The dog owner lives between Ken's and Leo's houses
   g. The bird is in the house with the street number higher than Ben's house.

Level Three

	BEN	RON	BOB	LEO	KEN	DOG	CAT	BIRD	PIG	GOAT
House #1										
House #3										
House #5										
House #7										
House #9										
Dog										
Cat										
Bird										
Pig										
Goat										

12. **The sum of Boris's and Rachel's ages** is 63 years. If Boris's age would be increased by 25% and Rachel's age would be increased by 20%, then the sum of their ages would be 77 years. What are the ages of Boris and Rachel?

Level Three

Level Three

LESSON 5

Classwork:

Roman Numerals (Symbols and Values): 1 = I; 5 = V; 10 = X, 50 = L; 100 = C; 500 = D; 1000 = M
Line on top of a number means times 1000.

1. Translate to Hindu-Arabic notation:

   a.  =

   b. =

2. Add/Subtract:

   a.

   b.

3. This year Sean is 17 and Dave is 29.
   a. When will Sean's age be half of David's age?
   b. When will Sean's age be one third of David's age?

4. Translate to Hindu-Arabic notation:

   a. MCDLXXIII =

   b. MMDXCVIII =

   c. CCCXXXIX =

   d. $\overline{\text{LXXXIV}}$ =

   e. $\overline{\text{XXMMMDLXXII}}$ =

   f. $\overline{\text{IVCCCXIV}}$ =

   g. $\overline{\overline{\text{DCCCLXXXVIII}}}$ =

5. Add/Subtract in Roman Numerals:
   a. CXXXIX + DCCCXXVII =
   b. CMLXXXV – CDLXIX =
   c. $\overline{\overline{\text{LII}}}\overline{\text{CDLVII}}$ + $\overline{\overline{\text{IV}}}\overline{\text{MMMCCCXXVI}}$ =
   d. CDXLII – CCCLXXXIV =
   e. MMMCCLVIII + MMDCCCXVI =

6. **Students in Level 2 of the Brain Power** program are allowed to move to a higher level (Level 3) if they show good progress in problem solving. Students of Level 3 move back to Level 2 if they do not do their homework and if their marks are low. In October, one student

moved up from Level 2 to Level 3. Now both classes have the same number of students. In November, one student from Level 3 moved back to Level 2. So, the number of students in these two classes became the same as it was before October. In December, another student from Level 3 moved back to Level 2. Now the number of students in Level 2 became twice the number of students in Level 3. How many students are in both classes?

7. **Bisector DM of an angle in rectangle ABCD** splits one side of the rectangle into two equal segments. What is the area of the quadrilateral DMBC if the width of this rectangle is 24 cm?

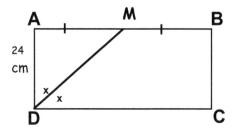

8. **Yaniv is 28 years old now and his** grandfather is 84.
   a. When was Yaniv's age one third of his grandfather's age?
   b. When was Yaniv's age one fifth of his grandfather's age?

9. **Last year Yaniv saved coins** (25c and 10c) in his piggy bank. When he opened his piggy bank this year, the total number of coins there was 232. The total value of all these coins was $52. How many quarters and how many dimes were there?

10. **The school's team participating in the swimming competition** consist of 7 boys and 2 girls. All the boys are the same age and all the girls are the same age. The total sum of their ages is 80 years. The team was split into two sub-teams: one sub-team consists of five boys and the rest of the boys and girls are in another sub-team. Now, the total sum of ages in each sub-team is the same. What are the ages of each boy and each girl?

11. **There are three Level 3 classes.** Today they are experimenting with logic problems. In the last problem, all students of one class decided to be truth-tellers, and another class's students decided to be liars. Now, all the students from these two classes are playing in the hallway, but they sent one of their students to the third class. The third class is solving a problem – they have to determine if this student (lets call her Student X) is a truth-teller or a liar. The problem is that the third class can only send Student X to the hallway to ask any one there if he/she is a liar or a truth-teller and bring back the answer. The answer brought back by Student X was that she met a liar in the hallway. Is Student X a truth-teller or a liar?

Fun Home Assignment

1. Write in Roman Numerals:

   a. 298 =

   b. 5687 =

   c. 35828 =

   d. 64325666 =

   e. 28999 =

   f. 7981324 =

2. Add/Subtract in Roman Numerals:

   a. MMMDLXXXVIII
      + MMCMLXXIX

   b. MMMDLXXXVIII
      − MMCMLXXIX

   c. CCCLXXXIII
      + DCCCXXVII

   d. DCCXXVII
      − CCCLXXXIII

   e. $\overline{\text{XIV}}$DCCXCIX
      + MMCCXXVI

   f. $\overline{\text{XIV}}$DCCXCIX
      − MMCCXXVI

3. **Four drunken men are looking at the table.** On the table is a lineup of a teacup, a coffee mug, a glass jar, and a wine glass. The men know that in one of these containers is cognac, in another is whiskey, in the third is tequila, and in the fourth is peach schnapps. The men already figured out that schnapps and cognac are not in the teacup. They also know that the glass jar does not contain whiskey nor tequila. As well, the wine glass does not contain whiskey nor schnapps. Can they figure out what drink is in each container if the coffee mug is beside the wine glass and the container with cognac?

4. **Three friends, Elina, Rebecca, and Darya** sent a donation cheque to the Kids Hospital for a total amount of $682. Elina's contribution to this cheque was $\frac{7}{10}$ of Darya's contribution, and Rebecca's contribution was $\frac{5}{7}$ of Elina's contribution. How much did each girl contribute?

Level Three

5. **Find five consecutive even numbers** so that if the first number is doubled, the second number is increased by 4, the third number is divided by 3, the fourth number is decreased by 7, and the fifth is divided by 5, the sum of the resulting numbers is 153.

6. **In parallelogram ABCD,** the height BM is bisecting the side AD. Find the length of each side of parallelogram ABCD if the perimeter of this parallelogram is 62 cm and the perimeter of triangle ABD is 43.4 cm.

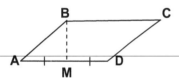

7. **Two diagonals in parallelogram ABCD** intersect in point O. Segment PT passes through point O in such a way that the length of segment BT is 8 meters and the length of segment AP is 9.2 meters. Find the length of side AD.

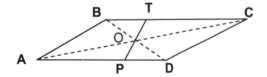

8. **The numerator of a fraction is** 5 less than the denominator. If 7 is added to both, numerator and denominator, the result is a fraction equivalent to $\frac{2}{3}$. What is the reciprocal to the original fraction?

9. **Five families living on the same** street made a deal with a contractor to renovate their houses. All of them need to replace windows and shingles on the roofs of their houses. Each family ordered a different type of window and a different colour of roof shingles. However, when the material was delivered, the neighbours discovered that the contractor mixed up their orders. Each house was delivered the material ordered by two different families not living in this house. Based on the clues below, please find whose material was delivered to each house:

   a. Lee's windows were delivered together with Perry's roof shingles

   b. Wong's roof shingles were delivered to the Jones's house

Level Three

c. Cohen's windows were delivered together with Jones's roof shingles

d. Cohen's windows and Jones's roof shingles were not delivered to Wong's house

e. Wong's windows were not delivered to the Perry's house

f. Lee's windows and Perry's roof shingles were not delivered to Wong's house

		Roof shingles					Windows				
		Jones	Cohen	Wong	Perry	Lee	Jones	Cohen	Wong	Perry	Lee
House	Jones										
	Cohen										
	Wong										
	Perry										
	Lee										
Windows	Jones										
	Cohen										
	Wong										
	Perry										
	Lee										

10. **Liran is 7 years older than his brother Idan.** Their mom's age is two more than three times the sum of Liran's and Idan's ages. How old are each of the family members if the sum of all three of their ages is 54?

11. **Fill in missing digits:**

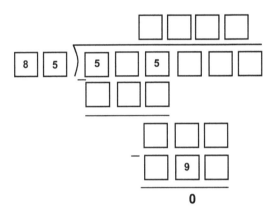

Level Three

Level Three

LESSON 6

Classwork:

1. **Solve the Conjunction riddles:**

   1.1 **First Riddle**
   a. Mrs. Cowen prepared special assignments for her three best students, Mike, Dana, and Elina.
   b. The three assignments include math, science, and English
   c. Mike did not want to do the English assignment, and Mrs. Cowen did not give him the science assignment
   d. Dana was ready to do any assignment, and Elina would not do science assignment.

   Mike's assignment is _____
   Dana's assignment is _____
   Elina's assignment is _____

   1.2 **Second Riddle:**
   Three boys, Jack, John, and Jake, planned to take their girlfriends, June, Julia, and Jane, to a concert. When asked, who went with whom to the concert, they gave three different answers:
   a. Jack invited June, and John invited Julia
   b. Jake invited Jane, and Jack invited Julia
   c. Jake invited June and John invited Jane

   It is known for sure, that only one of these answers is true. As well, it is a true fact that John broke up with his girlfriend and they both did not go to the concert. Which girl did not go to concert? Who went to the concert and with whom?

2. **Six scientists, Dr. Kohn, Dr. Brown, Dr. Cohen, Dr. Wang, Dr. Lin, and Dr. Rubin, from six scientific labs** visited various schools to give presentations about their research and about their fields of study. The pamphlet with their names, names of labs, and the schedule of their presentations was distributed to students in advance so students would prepare questions for the presenters. The information in the pamphlet includes the following:

- Dr. Kohn and presenter from the Robotics Lab are Mechanical Engineers
- Dr. Lin and presenter from the Laser Lab are mathematicians
- Dr. Wang and presenter from the Computer Technologies Lab are physicists
- Dr. Brown, Dr. Rubin, and presenter from the Bio Technology Lab attended a conference in Italy last year
- Presenter from the Computer Technologies Lab never travelled to Italy
- Presentation by the Chemistry Lab presenter is scheduled after the Dr. Kohn's presentation

Level Three

- Presentation by the Med Lab is scheduled after Dr. Wang's presentation
- Dr. Rubin is always presenting first
- Dr. Brown and presenter from the Robotics Lab studied at the same university
- Dr. Wang and presenter from the Chemistry Lab finished the same high school

Please find out the profession of each presenter and which lab he/she is representing

3. **There were 105 pirates on two ships** under the command of captains Crook and Huck. Last month the captains cooperated in capturing a large mercantile ship carrying gold coins. They split their booty equally between pirate ships. Each captain, then, split the coins equally among his pirates. How many pirates were on each pirate ship if each pirate on captain Crook's ship ended up with 8 coins and each pirate on captain Huck's ship ended up with 7 coins? How many gold coins did they capture from the mercantile ship in all?

4. **Sarah and Mike went out shopping** for a surprise birthday party for their mom. Each of them had the same amount of money in their pockets to spend. Sarah decided to buy medium quality chocolate candies for the price of $5.70 per pound and to pack these into a nice gift box. In addition, she paid $5.25 for the gift box. Mike did not care about packaging – he wanted his mom enjoy a good quality chocolate. He bought the same quantity of chocolate candies as Sarah but he paid $7.20 per pound for higher quality chocolate candies. How many pounds of chocolate candies did each of them buy?

5. **In parallelogram ABCD diagonals AC and BD** are equal in length. Prove that this parallelogram is a rectangle.

6. **In parallelogram ABCD diagonals AC and BD** are perpendicular to each other. Prove that this parallelogram is a rhombus.

7. **The length of each diagonal** in the square ABCD is 18 meters. A rectangle MNPQ is inscribed into the square ABCD in such a way that sides of the rectangle are parallel to the diagonals of the square and vertices of the rectangle are on the sides of the square. Find area of the rectangle MNPQ if its length is twice the width.

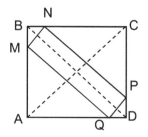

Level Three

8. **Solve equations:**

   a. $\dfrac{7}{X+3} = \dfrac{5}{2X-3}$

   b. $\dfrac{11}{3X-8} = \dfrac{2}{X+1}$

   c. $\dfrac{-4}{7Y+5} = \dfrac{9}{6Y+2}$

9. **A clock on the City Hall tower rings** 3 times at 3pm. This takes only 18 seconds. How many seconds will take this clock to ring 6 times at 6pm?

10. **Two mothers, Ann and Julia**, and their two daughters, Iris and Inga, went shopping in the music shop yesterday morning. The music shop had a great discount on classical music that day so all of them bought 35 CD's with classical music in all. Ann and her daughter Iris bought the same number of discs each. Julia bought two times more discs than her daughter Inga.

    a. How many CD's did each of them buy?
    b. How many different solutions does this problem have?

Level Three

Fun Home Assignment:

1. **Ann called her friend Veronica** to tell her about the great sale of classical music. Veronica, in turn, called Maria. In the afternoon, Veronica with her daughter and Maria with her daughter visited this music shop. These women also bought 35 CD's that afternoon. Veronica and her daughter Amy bought the same number of CD's. Maria bought three times more CD's than her daughter.

   a. How many CD's did each of them buy that afternoon?
   b. What is the name of Maria's daughter?

2. **Solve the Conjunction riddles:**

   2.1. **Four men, Mark, Max, John,** and Jack married their sweethearts at the City Hall last Sunday. The City Hall clerk processed the record of their marriages the next day and she was confused while trying to solve the puzzle. She received these four written messages:
   a. Mark married Dana, and Jack married Bonny
   b. John married Ann, and Max married Dana
   c. Jack married Sarah, and Mark married Ann
   d. Jack married Dana, and Mark married Ann

   The clerk knows that each written message has one true fact and one false fact. **Who married Sarah?**

   2.2. **Six students wrote test** on problem solving. Their marks were 65%, 79%, 83%, 88%, 92%, and 95%.
   . Anna's mark was 88% **and** Eddie's mark was higher.
   . Karolina's mark was lower than Mike's **and** higher than Emily's.
   . Mike's mark was higher than Anna's **and** lower than Eddie's mark.
   . Tina's mark was the lowest among these six students.
   What was each students' mark on the test?

3. **Mike's car is broken and is in a garage** for repair. The car needs special parts and it will take more than a week to get the car repaired. Meanwhile, every morning Mike is taking a taxi to his office. Yesterday, the taxi driver turned on the meter when Mike entered the taxi. There was an immediate charge of $3.50 and then Mike was charged $0.75 per each kilometer. This morning Mike stopped a different taxi. There was no immediate charge but Mike was charged $0.95 per kilometer. Mike was amazed when he was charged this morning exactly the same amount as he was charged yesterday. **How far is Mike's home from his office?**

Level Three

4. **Find such four consecutive even numbers** that if the first number is divided by 2, the second number is divided by 8, the third number is divided by 6, and the fourth number is divided by 4, the sum of the resulting numbers is equal to the third number among the consecutive even numbers.

5. **Prove:**
   a. **The point O of intersection** of diagonals in parallelogram ABCD is a mid-point of the diagonals.

   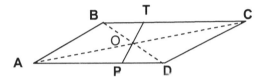

   b. **Line PT is passing through** point O. Prove that point O is a mid-point of segment PT

6. **Four students, Luka, Ariel, Rachel, and Mark** are suspected of cheating on the Math exam. When confronted by the school principal, Luka said that he never steals answers but Ariel was copying other students' answers. Luka also said that Mark is a truth-teller. Ariel insisted that he is innocent but it was Mark who copied other students' work. Mark said that Ariel is lying. Rachel insisted that she did not copy other students' work. The school principal figured out that only one of these four students is saying the truth. **Who copied other students work?**

7. **Triangle ABC is isosceles.** Find all possible values of X.

   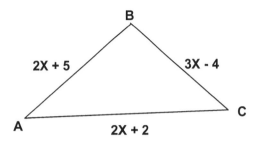

8. **Maayan is preparing for a new school year.** Yesterday she was buying pens and pencils for the whole school year. She spent exactly the same amount for all pens as she did for all pencils. The price of one pencil is 10 cents less than the price of one pen. With the money she had, Maayan bought 20 pens and 25 pencils. How much money did Maayan spend yesterday?

9. **The school supply warehouse received an order** for 67 binders and 41 paper pads. This order was assembled into three boxes, each box containing the same number of items. The number of binders in the first box was five times more than the number of paper pads. The number of binders in the second box was 14 more than the number of paper pads. How many binders and how many paper pads were packed in the third box?

Level Three

10. **There are four kinds of apples in the box**: yellow apples, green apples, red apples, and striped apples. You are blindfolded and cannot see the colour of the apples. As well, you cannot differentiate these apples by touching them as they have the same texture. What is the minimal number of apples you must take out while you are blindfolded, to be sure that:
   a. You have at least two apples of the same kind
   b. You have at least four apples of the same kind?

11. **Do you remember the story** of how Archimedes solved a problem, given to him by his king, about the golden crown? When he solved the problem, Archimedes ran with the solution to the king's palace yelling with great excitement, "**Eureka**"!
   Imagine that Archimedes is given a different problem:
   He is given 9 coins; one of these 9 coins is fake and is slightly lighter than the other coins (lets say by 1 milligram). Archimedes has to find this fake coin by using an old fashion "balance" scale without weights only 3 times. How can this be done?

12. **At the end of a hard days work,** five waiters in a "Very Yummy" restaurant, Jack, John, Jim, Jerry, and Bob, are comparing how much work each of them has done. They are counting how many soup and how many main course orders each of them served at lunch time. They were surprised to find out that each of the waiters served more soups than main courses. Please find out from the clues below how many soup and how many main course orders each of them served that lunch:

. Jim served more than 12 main course orders.

. The waiter, who served 26 soup orders, served 10 less main courses than the number of soup orders served by Bob.

. Jim served more soup orders than Jerry, but Jerry served more main courses than Jim.

. The waiter that served the highest number of main course orders, served 2 less soup orders than Jack

. Jack served 12 main course orders less than soups

. John served the highest number of main course orders

	Soup orders					Main Course				
	22	24	26	28	30	12	14	18	20	24
Jack										
John										
Jim										
Jerry										
Bob										

LESSON 7

Classwork:

1. Solve the Conjunction riddle (BUT):

   Four friends, Kathy, Karen, Ken, and Kim, are planning their summer vacations. Their choices are France, Greece, Russia, and Italy. Each of them chose to travel to a different country. Please figure out where each of the friends will travel if:

   . Kathy likes Russian literature **BUT** she decided not to go to Russia.

   . Ken was planning to travel to one of Mediterranean countries **BUT** he decided not to go to France or Greece.

   . Karen invited Kim to travel together to France **BUT** Kim decided to go to Russia

2. In each of the following Exclusive "OR" sentences one part is true and another part is false. Solve the following disjunction ($\underline{V}$) riddles:

   a. **Find out the unique profession of each person:**
      . Adina is a physician or a painter
      . Arie is a computer programmer or he is an engineer
      . Ann is a computer programmer or she is a physician
      . Arnold is a painter or an engineer
      . Adina is a painter or a computer programmer

   b. **The following information was available from the school athletic competitions:**
      . Molly won the 100 meters dash event or Maria won the long jump event
      . Mike won the high jump event or Mitch won the 100 meters with hurdles
      . Molly won the high jump event or Mike won the long jump event
      . Two boys won the running events

      Who won the 100 meters dash event?
      Who won the high jump event?
      Who won the long jump event?
      Who won the 100 meters with hurdles event?

   c. **After a long ballroom dance competition**, only four couples remained on the floor. Please find out who was partnered with Elaine, if:
      . Boris danced with Ada or Ben danced with Ada

      . Emma danced with Peter or Emma danced with Mike

      . Boris danced with Ann or Peter danced with Ann

      . Ada danced with Boris or Ada danced with Mike

Level Three

3. Jerry, Jacob, and John won $1175 in a lottery. Now they need to split this prize among themselves according to the contribution each person made towards buying the lottery ticket. How much should each of the friends get if Jacob's contribution was $\frac{2}{3}$ of Jerry's contribution and John's contribution was $\frac{6}{7}$ of Jacob's contribution?

4. The Bank's branch at Main Street has 3 less tellers than the branch on Second Street. The Main Street branch manager complained that more clients are visiting her branch but they do not have enough tellers to deal with work pressure. The Bank's VP of the area solved the problem by transferring six (6) tellers from the branch on Second Street to the branch on Main Street. As a result of this transfer, the number of tellers in the branch on Second Street became $\frac{2}{5}$ of the number of tellers in the branch on Main Street. How many tellers are now in each branch?

5. In the quadrilateral ABCD drawn diagonal BD, splitting the quadrilateral into two triangles. Based on information provided in the figure below, please sort out sides of the quadrilateral, AB, BC, CD, AD from shortest to longest:

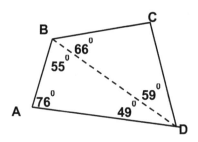

6. Only five students in Mrs. Cowan's class got marks on the latest math test lower than 80. Each of these five students, Molly, Maria, Mike, Morton, and Melisa, did not solve problems well on one of the five particular topics: integers, exponents, radicals, BEDMAS, and geometry. Read the clues below and find out what mark each of these five students received on the test and what topic each of them will have to review before the next test:

. Molly's mark was higher than Mike's, but two points lower than the mark of the boy who made mistakes in Order of Operations (BEDMAS)
. Maria knows exponents better than integers
. A girl that made mistakes in exponents received one point more on the test than the girl who made mistakes in integers
. The mark of the boy who made mistakes in geometry, is lower than the mark received by the girl who

Level Three

made mistakes in radicals.

	Marks					Integers	Exponents	Radicals	BEDMAS	Geometry
	65	68	70	72	73					
Molly										
Maria										
Mike										
Morton										
Melisa										
Integers										
Exponent										
Radicals										
BEDMAS										
Geometry										

7. **The quadrilateral ABCD is a trapezoid**: BC ∥ AD. Point E splits side CD into two equal segments CE = ED. Prove that the area of the △ABF = area of ABCD.

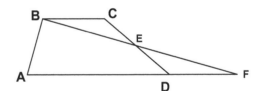

8. Fill in the missing numbers:

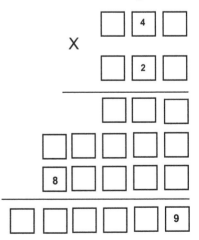

9. **Four students decided to buy a gift** for their classroom teacher on her birthday. They decided to split the cost equally. However, three more students expressed desire to participate in buying the gift and they were willing to pay their share of the gift's price. The original four students agreed to this offer as they realised that this will save each of them $12. What is the price of the gift?

10. Fill in the missing numbers:

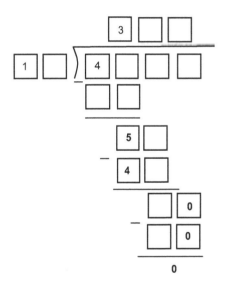

Level Three

Fun Home Assignment:

1. **Solve the "BUT" riddle:**

   Four women, Nona, Nicole, Nina, and Nora are married to Rob, Ron, Rick, and Roger (but not necessarily in this order). Find out who is married to whom if:

   . Nona thought that Ron is Nina's husband **BUT** Ron is not married to Nina

   . Rick and Roger were Nicole's teachers in high school **BUT** Nicole's husband does not know them

   . Ron played golf with Nicole's husband **BUT** Roger never met Nona's husband.

2. **Solve "OR" riddle (Exclusive Disjunction):**

   Five girls, Betty, Becky, Beata, Bonny, and Bluma, and five boys, Alex, Adam, Alon, Arie, and Anton are working on five projects: Earth, Pollution, Demographics, Medicine, and Sports. One girl and one boy are working together on each project. Find out which pair of boy and girl are working on which project if:

   . Betty and Adam **OR** Becky and Anton work on the Demographics project

   . Beata and Adam **OR** Bluma and Anton work on the Pollution project

   . Bluma and Arie **OR** Betty and Adam work on the Demographics project

   . Bluma and Arie **OR** Beata and Alex work on the Sports project

   . Beata and Alex **OR** Becky and Arie work on the Earth project

3. Solve equations:

   a. $\dfrac{3}{2X-7} = \dfrac{5}{4X+2}$

   b. $\dfrac{3X-7}{2X+5} = \dfrac{6X+1}{4X-2}$

   c. $\dfrac{4X+5}{2X-3} = \dfrac{8X+9}{4X+5}$

4. **The Summer Bikini store sold** popular women's swimming suits for the total amount of $13448 in April. The most expensive suits were sold for $44 per piece, the second category suits were sold for $33 per piece, and the least expensive were sold for $32 per piece. The number of the most expensive suits sold was $\dfrac{3}{4}$ of the number of the second category suits sold. The number of the cheapest swimming suits sold was

$\frac{2}{3}$ of the number of the most expensive swimming suits sold. How many swimming suits in total were sold in April?

5. **Liran has $3 less than double the amount Lior has.** After Lior added to his amount another $\frac{1}{2}$ of the money he started with, he still has $8 less than Liran. How much money do Lior and Liran have altogether?

6. **Angle A is 12⁰.** It is "isosceable" because of the following design:

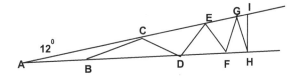

AB = BC = CD = DE = EF = FG = GH = HI.

  a. Prove that ∠AIH = ∠AHI; i.e. △ AIH is isosceles
  b. Find a few more "isosceable" angles measured in integer degrees.
  c. How many "isosceable" angles are there that are measured by integer degrees?

7. Joseph and Julia helped their uncle Donald on his farm during their summer vacations. At the end, uncle Donald gave them $720 in total for their work and he divided these $720 into two amounts based on the number of hours Joseph and Julia worked. As a result, Julia's amount was much larger than Joseph's. Joseph, not so happy with the amount he received, asked Julia to give him as much money as he received from uncle Donald. Julia, being a good sister, gave Joseph as much money as he received from their uncle. But now Julia's amount became too small and she asked Joseph to give her back as much money as the amount she was left with. Joseph complied with Julia's request, but his amount remained too low after he gave money to Julia. He asked Julia again to give him as much as he was left with after the latest transaction. This time, Julia became unhappy again and she asked Joseph to give her back as much as she remained with. To their surprise, they each came up with the equal amount of $360 after the latest transaction. How much money did uncle Donald give to Joseph and how much money did he give to Julia?

8. The least of four consecutive even numbers is divided by 2, the second is divided by 3, the third is divided by 2, and the fourth is divided by 4. Find these four consecutive even numbers if the sum of the quotients is 39.

Level Three

9. **Find the product of the exercise below:**

$(1 + \frac{1}{2})(1 + \frac{1}{3})(1+ \frac{1}{4})(1 + \frac{1}{5})\ldots(1 + \frac{1}{99})(1 + \frac{1}{100}) =$

10. **Fill up all the missing numbers:**

How many different solutions are there?

Level Three

LESSON 8

Classwork:

1. **Mr. Donald's car suddenly stopped in the middle of the road.** No matter how many times Mr. Donald tried to restart the engine, there was no success. Mr. Donald decided to call his garage mechanic. After Mr. Donald explained on the phone what had happened to the car, the car mechanic said that it could be a problem with the ignition or with the battery. The car was towed to the garage. Two hours later Mr. Donald's car was repaired and he paid handsomely for a new battery and for a new set of spark plugs.
   a. Was the diagnose made by the car mechanic correct?
   b. Would the diagnose be correct if the problem was only with the battery?
   c. Would the diagnose be correct if the problem was only with the ignition?
   d. Would the diagnose be correct if the problem was because the gas tank was empty?

2. Leo is telling his brother Sam about the interesting occurrence his friend Rebecca had two weeks ago. She was walking around the block one evening and she found an envelope on the ground. Rebecca opened the envelope and saw money inside. Leo does not remember exactly, but he thinks Rebecca told him that there was a $100 bill or a $50 bill in the envelope. "What did she do with this envelope?" – asked Sam. "She went to the police station and gave them the money she found" – said Leo. "Hmmm", said Sam, "This could be exactly the envelope I lost two weeks ago, but I had only $50 in my envelope. Let's call Rebecca to check!". On the telephone, Rebecca said that there were two bills in the envelope, a $100 bill and a $50 bill.
Did Leo pass the correct information to Sam about the content of the envelope found by Rebecca?

3. Given △ ABC is isosceles. Sides AB and BC are equal in length. ∡ DBA is external. Line EB is a bisector of ∡ DBA. Prove that EB ∥ AC.

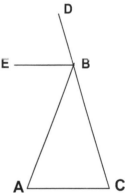

4. **One side of ∡ A is split into segments** AB_1, B_1B_2, and B_2B_3. The other side is split into segments AC_1, C_1C_2, and C_2C_3. It is given that $B_1C_1 \parallel B_2C_2 \parallel B_3C_3$. Find the length of AB_3 and AC_3 if $AB_1 = 3m$, $B_1B_2 = 4.5m$, and $AC_1 = 5.4m$, and $C_2C_3 = 7.5m$.

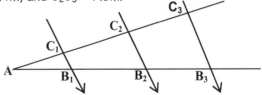

Level Three

5. **Four students, Glen, George, Greg, and Gabriel** are bragging about their marks on the last math test. They got 83%, 87%, 90%, and 94% on the test. Glen said that his mark was not the best and not the lowest among these four students. George said that his mark was not the lowest among these four students. Greg said that his mark was the best and Gabriel said that his mark was the lowest. One of these students is not telling the truth. Figure out:
   a. Who of these four students lied?
   b. Who's mark is the lowest?
   c. Who's mark is the highest?

6. **The numerator of a fraction is 4 less** than the denominator. If 5 is added to both numerator and denominator, the new fraction will be equivalent to $\frac{2}{3}$. What is the original fraction?

7. **Joseph and Robert want to buy** a basketball. They counted their money and found out that the total amount is $36.65. This is $0.85 lower than the price of the basketball. The boys have 137 coins consisting of $1 coins, quarters, and dimes. How many quarters and how many dimes do they have if the number of quarters is 3 times larger than the number of $1 coins?

8. **Translate to base 10:**

   $12_4 =$ $12_6 =$

   $23_5 =$ $22_3 =$

9. **Translate from base 10:**

   $12_{10} = \_\_\__4$ $14_{10} = \_\_\__6$

   $27_{10} = \_\_\_{12}$ $35_{10} = \_\_\__8$

10. **Four friends, John, Jordan, Jeremy, and Jim** work at a factory that manufactures pianos. Their job is to deliver pianos to stores. It requires 3 men to move one piano. Yesterday, John moved 13 pianos; he moved more pianos than any of his co-workers. Jordan moved 10 pianos; he moved less than any of his co-workers. How many pianos did all of them move yesterday altogether?

Level Three

Fun Home Assignment:

1. **Translate to base 10:**

   a. $22_7 = \underline{}_{10}$ $31_4 = \underline{}_{10}$

   b. $17_8 = \underline{}_{10}$ $23_{12} = \underline{}_{10}$

   c. $43_6 = \underline{}_{10}$ $33_5 = \underline{}_{10}$

   d. $22_3 = \underline{}_{10}$ $100_2 = \underline{}_{10}$

2. **Translate from base 10:**

   a. $9_{10} = \underline{}_4$ $17_{10} = \underline{}_8$

   b. $18_{10} = \underline{}_{11}$ $33_{10} = \underline{}_7$

   c. $26_{10} = \underline{}_5$ $66_{10} = \underline{}_9$

   d. $49_{10} = \underline{}_7$ $35_{10} = \underline{}_6$

3. **Three friends, Kathy, Dasha, and Elina, collect classical music on discs.** If Kathy would have 5 more discs, then she would have as many discs as the other two girls together. If Dasha would have 9 more discs, then she would have as many discs as the other two girls together. Figure out how many discs each girl collected and what each girl's last name is if:

   . The number of discs collected by McKay is a multiple of 3

   . Johnston collected 11 discs

   . Cohen is the best student in math in her class

4. $\triangle$ABC is isosceles: AB = BC. Segment AD is bisecting $\angle$ A. Given that AD = BD = AC. Base AC is extended and a segment EF is drawn through point D in such a way that EF $\perp$ AB.
   a. Find values of all angles in $\triangle$ ABC
   b. Prove that $\angle$ B = 2·$\angle$ F

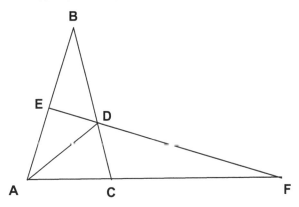

5. **Rachel had $2 more than four times the amount Henry had.** When Rachel gave Henry $6, Rachel's amount became twice larger than Henry's. How much did each of them have at first?

6. **One quarter of the original number is increased** by $\frac{4}{7}$ of a number that is two (2) larger than the original number. The result is one less than the original number. What is the original number?

7. **Prove that the mid-line in any triangle** is parallel to the base (line across the mid-line) and is equal to half of the length of the base. (Use

Level Three

rules of similarity and congruency of triangles)

8. △ ABC is isosceles: AB = BC. The length of the base AC = 20m. Vertices E and F of the square DEFG are on sides AB and BC. Side DG is on the base AC and DG = 12m. Find area of △ ABC

9. **There are 35 problems on Brain Power's IQ Test.** Students get 10 point for every problem solved correctly. Students' mark is reduced by 7 points for every problem solved incorrectly. Students get 0 points for every problem they skip. Liran's mark on this test was only 17 points. How many problems did Liran skip on this test?

10. **Joe-Hard Fist, a chief of a band of thieves,** wanted to celebrate his birthday. He likes rabbit stew and he wanted to have it for this occasion. Joe-Hard Fist called two old members of his gang, Gook and Crook, and he ordered them to get three white rabbits from Mr. Donald's farm. He knows that Mr. Donald is growing rabbits and that there are 24 white rabbits, 15 grey rabbits, and 21 brown rabbits on his farm. There was no moon and no stars at night when Gook and Crook snuck into the rabbit pen at Mr. Donald's farm. It was so dark that Gook and Crook could not see the colour of the rabbits. They started to think about what is the minimum number of rabbits that they would have to take with them from Mr. Donald's farm so there for sure would be at least three white rabbits. They could not figure this out but they knew that the number is not small. They came to the farm with a small cage that could fit up to 10 rabbits but this would not be enough. They returned to Joe-Hard Fist and explained him their dilemma. "OK", said Joe-Hard Fist, I can settle for three rabbits of any colour.

  a. What is the minimal number of rabbits that Gook and Crook would have to take from Mr. Donald's farm to be sure that they have at least 3 white rabbits?

  b. What is the minimal number of rabbits that Gook and Crook would have to take from Mr. Donald's farm to be sure that they have at least 3 rabbits of the same colour?

11. **Eden is 6 years older than Tal.** Ruth is three times older than Eden. The sum of their ages is 10 more than seven times Tal's age. How old is each of them?

Level Three

LESSON 9

Classwork:

1. **Translate the following English sentences** into Boolean expressions:
   a. Tonight Tom has little time to do his homework. He will do his Math work OR both, English assignment and his Science project.

   b. Tonight Tom has little time to do his homework. He will do either both his Math work and English assignment OR his Science project.

   c. Tonight Tom has little time to do his homework. He will do his Math work AND either his English assignment or his Science project.

   Are the sentences conveying the same meaning? Can you rewrite these sentences to prove your point?

2. **A police inspector received a phone call** from a bank robbery witness. The witness reported that he saw both, Jack and Bully in the bank at 11 am. The inspector wrote remarks in his journal that this testimony is not true as he received verifiable info that Jack or Bully had been playing golf at that time.

   Translate this situation into a Boolean expression using De Morgan's law.

3. "Mister Peterson, how old is your granddaughter Fay?" asked Anna. Mr. Peterson answered that the number of months of his granddaughter's age is the same as his own age in years. "How old are you, Mr. Peterson?" asked Anna again. Mr. Peterson answered that his age together with his granddaughter's age in years equals 65.

   How old is Fay?

4. Translate to base 10:

   $2643_7 = $ ____ $_{10}$ $21022_3 = $ ____ $_{10}$

   $5614_8 = $ ____ $_{10}$ $101101001_2 = $ ___$_{10}$

5. Prove:

   $$\frac{2}{x+4} - \frac{2}{x+6} > 0 \text{ ; if } x > 0$$

6. Given two consecutive numbers; the first of these numbers is odd and the second is even. The sum of the odd number with four times the consecutive even number is no more than 71.

   What is the largest odd number that satisfies this condition?

Level Three

7. **ABCD is a parallelogram.** Mid-points of the sides of ABCD (MNOP) are connected.
   a. Prove that MNOP is also a parallelogram.
   b. Find area of MNOP if area of ABCD is 92 cm².

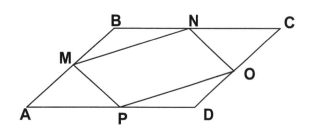

8. **A con woman Gigi bought a very nice blouse** in Mr. Smith's clothing store. She paid $68.50 for the blouse. The next day she came back to the store to exchange the blouse. She claimed that the blouse was a bit short and the blouse's colours were not exactly matching her skirts. This time Gigi chose a night dress which was priced at $137. She took the dress and left the store without paying. Mr. Smith immediately called the police and Gigi was arrested shortly. Police questioned her about why she didn't pay for the dress? She explained that she just returned the blouse which costs $68.50. Yesterday she also paid $68.50. Altogether it makes $137 which is exactly the price of the dress she took from the store today. So, she said, I did not have to pay extra money for the dress!
   What do you think will happen to Gigi?

9. **Solve equations and inequalities:**

   a. $|3X - 7| = 8$ $5 - |2.5X + 4| = 2$

   b. $|3X - 7| \geq 8$ $5 - |2.5X + 4| < 2$

10. **Rectangle ABCD has sides 6m and 10m.** On three sides, AB, BC, and CD there are semicircles built. Another rectangle, MNOP, has three sides, MN, NO, and OP, tangents to the semicircles, and the fourth side, MP, continuation of the side AD.

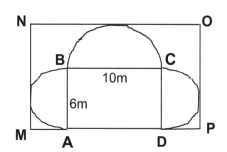

What is the area of the rectangle MNOP?

11. **Fill in the missing numbers:**

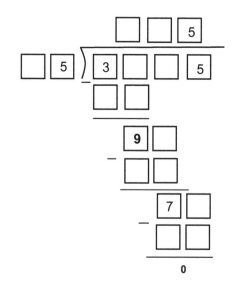

Level Three

Fun Home Assignment:

1. **Four students, Elliott, Eliran, Ethan, and Emily** decided to spend three weeks of their summer vacations doing some voluntary work. For the first week, they chose to visit sick elderly people, collect food for poor families, clean apartments for old and sick people, and do some shopping for disabled people. For the second week, they decided to volunteer at a Sick Kids hospital, at a Family Physician's office, at the Eye Doctor's Clinic, and at a General Hospital. For the third week, they decided to work at the Red Cross office, at a Food Bank warehouse, at a local church, and in the Rehab Clinic. Please find out what volunteering work each of the students did every week based on the clues below:
   - The student that worked at a Family Physician's office was working at a Rehab Clinic on the third week
   - The student that worked at a local church on the third week, collected food for poor families on the first week
   - The student that visited sick elderly people on the first week worked as well at a Sick Kids hospital and at a Food Bank warehouse
   - Elliott did not work in the Red Cross office
   - Ethan did not volunteer at a General Hospital
   - Ethan and Emily did not clean apartments for old and sick people; they also did not do shopping for disabled people
   - The student that worked in the Eye Doctor's Clinic also shopped for disabled people

2. **Translate to base 10:**

   a. 22010_3 = ____$_{10}$ b. 11321_4 = ____$_{10}$

   c. 43212_5 = ____$_{10}$ d. 43212_6 = ____$_{10}$

   e. 54321_7 = ____$_{10}$ f. 2817_9 = ____$_{10}$

3. **Add/Subtract in the given bases** (do not translate to base 10!)

   a. 22010_3 b. 11321_4 c. 43212_5
      $+ 12121_3$ $- 10332_4$ $+ 44243_5$

   d. 43212_6 e. 54321_7 f. 2817_9
      $- 43124_6$ $+ 44546_7$ $- 1828_9$

4. **A group of eight students decided to earn money** during their summer vacations. These eight students found work as farmhands at Mr. Smith's farm. Each of them would earn the same hourly pay and each student's pay for an hour's work will be in a whole

Level Three

number of dollars (no cents). After the first 5 hours of work, the total amount they earned altogether was less than $380. However, after the first 10 hours of

work, the total amount they earned altogether was more than $700.

What was each student's hourly pay?

5. **Solve the following equations and inequalities:**

   a. $\dfrac{3x-7}{4x+3} = \dfrac{6x+1}{8x-5}$

   b. $|8x+9| \leq 7$

   (Illustrate your answer on the number line)

   c. $\dfrac{5}{6x+1} - \dfrac{1}{2} = \dfrac{x-1}{3-2x}$

   d. $5(x-1) > 29 - 4x$

   (What is the lowest whole odd number that satisfies this inequality?)

   e. $\dfrac{7x-2}{3} - \dfrac{x}{6} > \dfrac{2x}{5} - \dfrac{4x+3}{2}$

6. **Sum of an even number** and five times its consecutive even number is more than 92. What is the lowest even number that satisfies this condition?

7. **Place single or multiple arithmetic operations (+, -, ÷, ×) and brackets** between these numbers to make results equal to 1. Can you find out an algorithm (an on-going process) that will allow having the sum in any row being equal to 1?

   1 2 3 = 1
   1 2 3 4 = 1
   1 2 3 4 5 = 1
   1 2 3 4 5 6 = 1
   1 2 3 4 5 6 7 = 1
   1 2 3 4 5 6 7 8 = 1
   1 2 3 4 5 6 7 8 9 = 1
   1 2 3 4 5 6 7 8 9 10 = 1
   1 2 3 4 5 6 7 8 9 10 11 = 1
   ...

8. Two legs AB and AC of an isosceles right triangle ABC equal 15 meters each. Rectangle AMNP is inscribed in the triangle ABC so ∠A is common to both the triangle and the rectangle. The length of side AP is 13 meters. The same way rectangle MQRS is inscribed into the triangle MBN. The side MS of this rectangle is 9 meters long. What is the area of rectangle MQRS?

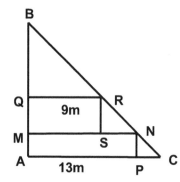

Level Three

9. **In September, Language Arts** students purchased 6 tickets to a Debaters show and 40 tickets to the Hamlet show for the total price of $472. In October they bought 12 tickets to a Debaters show and 37 tickets to the Hamlet show for the total amount of $514. The prices for the tickets in September and October were the same. What would be the total cost of 7 tickets to a Debaters show and 15 tickets to the Hamlet show?

10. **The "Fresh Fruit and Vegetables"** store owner was happy yesterday as he made a record sale. Yesterday morning he filled up his top shelf with eggplants, the middle shelf with cucumbers, and the bottom shelf with tomatoes. He counted 675 items in all on all three shelves. During the day he sold half ($\frac{1}{2}$) of the eggplants, two thirds ($\frac{2}{3}$) of the cucumbers, and three quarters ($\frac{3}{4}$) of the tomatoes. He was amazed to find that the number of items on each shelf was the same at the end of the day.

How many of each kind of vegetables were on the shelves yesterday morning?

11. **Fill in the missing numbers:**

        ☐☐☐
    X   ☐☐☐
    ─────────
       ☐0☐8
      ☐☐5☐
    ☐1☐☐☐
    ─────────
    ☐☐☐☐☐

12. **Perimeter of a large square** is four times the perimeter of another smaller square. How many times is the area of the larger square more than the area of the smaller square?

Level Three

Level Three

LESSON 10

Classwork:

1. Build a truth table for the sentence:

   IF it will be snowing whole day **Then** classes will be cancelled.

   Compare the truth table of the above sentence to the truth table for the sentence below:

   IF classes will not be cancelled **Then** it will not be snowing today.

   What is your conclusion, are these sentences equivalent?

2. Prove:

   $a \rightarrow b = \bar{a} \lor b$

3. **Analyse the following four implications.** What theorem do these implications prove?

   a. If there will be enough students that sign up for the ski classes, then our class will spend a week at a ski resort

   b. If there will be enough snow then the ski resort will be open for the season

   c. If the ski resort will be open then most of the students will sign up for the ski classes

   d. If there will be no snow in the resort area then the trip will be cancelled

4. Four students, Elizabeth, Eddie, Eden, and Ernie are playing in a ping pong tournament: each student plays one game against every other student. In ping pong there are no ties! The players agreed that for every win a player will get 2 points and for every loss a player only gets one point. At the end of the tournament, Eddie scored 6 points, Elizabeth scored 5 points, and Eden scored 4 points.
   a. How many points did Ernie score?
   b. Who won which game?

5. Given quadrilateral ABCD. Side AB is 3 meters longer than half of side BC. Side AD is 5 m shorter than twice the side BC. Side CD is twice longer than side AB. Find length of each side of this quadrilateral if its perimeter is 40 meters.

6. Translate to base 10:

   a. $B5A3_{12}$ = _____ $_{10}$

   b. $A2CD_{14}$ = _____ $_{10}$

   c. $4FDA_{16}$ = _____ $_{10}$

   d. 11000100111_2 = _____ $_{10}$

Level Three

7. Add/Subtract:

a. $B3A5_{12}$
   $+ 98B9_{12}$

b. $5F8C_{16}$
   $+ E97D_{16}$

c. $B3A5_{12}$
   $- 98B9_{12}$

d. 1101100001_2
   $- 1010110101_2$

8. Four boys, Joe, Daniel, Robert, and Jim are from Toronto, Ottawa, Montreal, and Calgary. Please figure out who is from which city if:

a. Jim and a boy from Toronto like playing chess with each other via Internet

b. Joe and a boy from Calgary are relatives. They never travelled to Toronto, nor to Montreal

c. Robert and a boy from Toronto are exchanging difficult math problems. They like to solve these problems together but they do not like when the boy from Calgary is trying to join them when they work on a math problem.

9. Triangle ABC is equilateral; each side is 20cm long. What is the length of the radius of the circle inscribed into this triangle?

10. Prove:

If $(2X - 3)^2 < (4X + 5)(X-1)$, then $X > 1\frac{1}{13}$

11. At the end of her shift, a cashier is counting money left in her cash register. She sorted out $20 bills, $10 bills, $5 bills, and $2 coins (toonies), $1 coins (loonies), and quarters. Her total amount is $665.50. How many coins did she count in all if the number of $20 bills is 2 less than the number of $10 bills, the number of $5 bills is 5 less than the number of $10 bills, the number of toonies is one less than twice the number of $5 bills, the number of loonies is 5 more than twice the number of $20 bills, and the number of quarters is 6 more than three times the number of $20 bills.

12. Next year Nicole's age will be half of her father's age. Ten years ago Nicole's father's age was 5 less than three times Nicole's age then. How old are they now?

Level Three

Fun Home Assignment:

1. **Mrs. Shohat is working at a travel agency.** Today she received a complex order from her best client. The client wants to maximize travel during her summer vacations. She is a teacher and summer for her is the only time for long travels. This summer, Mrs. Shohat's client wants to visit Toronto, New York, Ottawa, Washington D.C., and Boston. She also wants to use different means of transportation for her trips: car, train, bus, plane, and helicopter. In each city the client will travel to, she wants to visit one of the following: museums, art galleries, monuments, suburbs, and downtown. Mrs. Shohat is trying to make the travel arrangements based on the pieces of information her client provided her with:
   - The client does not want to visit downtown on her trip to Boston
   - A trip to Toronto must be by car but the client does not want to see monuments there
   - A trip to Washington D.C. should not be made by train. The client wants to visit suburbs there
   - A trip to Ottawa should not be made by plane nor by bus
   - On a trip by plane, the client wants to visit art galleries
   - On a trip by bus, the client wants to visit downtown

   Please help Mrs. Shohat make travel arrangements for her client.

2. **Translate to base 10:**

   a. $A5A4_{11}$ = _____$_{10}$

   b. 7777_8 = _____$_{10}$

   c. $2FC9_{16}$ = _____$_{10}$

   d. 9999_{12} = _____$_{10}$

   e. $DBA8_{14}$ = _____$_{10}$

   f. $55B5_{13}$ = _____$_{10}$

3. **Calculate:**

   a. $\quad 5E38_{16} \quad\quad 5E38_{16}$
      $+ \underline{37FB_{16}} \quad - \underline{37FB_{16}}$

   b. $\quad 80AD_{14} \quad\quad 80AD_{14}$
      $+ \underline{7CB9_{14}} \quad - \underline{7CB9_{14}}$

   c. $\quad 11101100100011_2$
      $- \underline{10110110101101_2}$

   d. $\quad 11101100100011_2$
      $+ \underline{10110110101101_2}$

   e. $\quad 432E_{15} \quad\quad 432E_{15}$
      $+ \underline{2D34_{15}} \quad - \underline{2D34_{15}}$

4. **Given parallelogram ABCD.** Mid points of the four sides of this parallelogram, M, N, O, and P are connected to form a smaller

Level Three

parallelogram MNOP. Through the vertices of the parallelogram ABCD are lines drawn parallel to diagonals of ABCD.

a. Prove that MNOP is a parallelogram.
b. Prove that EFTR is a parallelogram
c. What is the ratio of the area of parallelogram MNOP to the area of parallelogram EFTR?

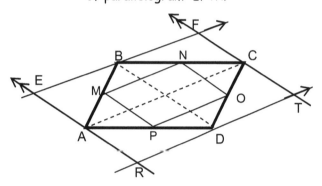

5. **Trapezoid ABCD is isosceles**; AB =CD. Base BC = 17m and base AD = 23m. Find the length of sides AB and CD if the area of the trapezoid equals 60 m².

6. **Farmer Smith is proud of his record making brown cow.** This cow produces the sweetest and the tastiest milk in the whole region. However, this fall the cow was sick and lost 25% of her weight. Mr. Smith put the cow on a special diet prescribed by the local veterinarian and the cow gained back 20% of its weight in one week. The next week the cow lost 10% of her weight again but a week later it gained 20% again. Did the cow regain her weight? Does the cow now weigh more or less than her weight before it become sick? By how much?

7. **After all, Mr. Smith decided to sell his cow.** He sold it to his neighbor Mr. White for $3630. Mr. White paid in cash and he put all the money for this cow in an envelope. Mr. Smith counted the money received and he sorted out the money based on the bills' denomination. There were $100 bills, $50 bills, $20 bills, $10 bills, and $5 bills in the envelope. The number of $50 bills was 3 more than 5 times the number of $100 bills. The number of $20 bills was 5 more than half of the number of $50 bills. The number of $10 bills was one more than 7 times the number of $100 bills. The number of $5 bills was 3 more than the number of $100 bills.
How many bills were in the envelope received by Mr. Smith?

8. **Adam is 8 years older than Eve.** 17 years ago Adam was twice older than Eve. In how many years will Adam's age be twice Eve's age now?

9. **In these three Fibonacci patterns a few of the terms are missing. Please fill in the blanks:**

a. 5, ___, ___, 11

b. 3, ___, ___, ___, ___, -1

c. ___, 2, ___, 10

10. Fill in all the missing digits:

```
  x  [2][ ][4][ ][8][ ]
           [ ][ ][ ][5]
         ─────────────
        [ ][ ][7][3][ ][2][ ]
     [2][ ][ ][ ][8][ ][ ]
  [ ][6][ ][ ][7][0][ ][ ]
[7][ ][ ][ ][ ][ ][ ][ ]
─────────────────────────
[ ][ ][ ][ ][ ][ ][7][ ][5]
```

11. Solve:

  a. |X - 1| = |X - 2|

  b. | 4X - 7 | > 3

Level Three

Level Three

LESSON 11

Classwork:

1. Translate from Base 10:

   a. $54_{10} = \underline{}_3$

   b. $75_{10} = \underline{}_6$

   c. $33_{10} = \underline{}_4$

   d. $28_{10} = \underline{}_2$

   e. $165_{10} = \underline{}_{14}$

   f. $3617_{10} = \underline{}_{16}$

2. **The rumors were that three students from Brain Power Enrichment team** won the regional competition in problem solving. George is trying to guess which of these students it could be, but he only managed to find out their last names: Cowan, Levy, and Garcia. George's friend, Tim, thinks that the winners' first names are Laura, Fred, and Jonathan. "Aha", said George, "Laura's last name is Cowan". "This cannot be true", said Tim. "I know this because Marina Berman told me last week that Marina's father and Levy's mother are twins. She told me that her father, Levy's mother, and Jonathan's grandfather, Mr. Gordon, are planning a party for our best problem solvers. They are planning to give 3 different gifts cards for the winners: $100 card, $70 card, and $50 card. I heard that Jonathan's gift card will cost $20 less than Cowan's card". **Which gift card will Laura receive? What is the full name of each winner?**

3. **Three friends, Lev, Yonatan, and Gleb** worked at the farm last summer. Lev earned $1200 for the whole summer. Yonatan earned as much as Lev and half of the amount earned by Gleb. Gleb earned as much as both Lev and Yonatan together last summer. How much money did all three friends earn last summer altogether?

4. **The difference of two numbers is 13.** If five times the smaller number is subtracted from four times the larger number, the difference is 31. What is the sum of the two original numbers?

5. In △ABC, AM = MC and CN = NB. Perimeter of △ABC is 72 cm. Find the length of the segment MN if
   a) AB is 5 cm shorter than AC
   b) Twice AC is 7 cm shorter than three times BC.

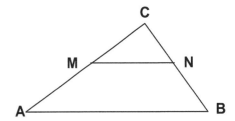

Level Three

6. In △ABC side AC = 36 m, side AB = 28 m, and side BC = 24 m. Points M and N are mid-points of the respective sides AB and BC. The line connecting points M and N is extended from point N to the point P. It is given that the length of NP equals half of MN. Segments PR and CT are drawn parallel to AB. Segment QT is parallel to AC.
What is the perimeter of the quadrilateral QTCR?

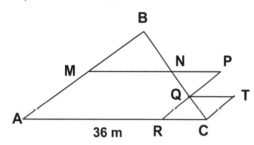

7. Solve the equation and explain the result:

$8X + 3(X+2) = 15 + 11x$

8. **Last summer Ron travelled across Canada in his car.** In the first two days he made $\frac{1}{5}$ of the whole trip. In the next 3 days he covered $\frac{3}{4}$ of the remaining distance. In the sixth day he made only 90 kilometers. The remaining distance, which was $\frac{7}{40}$ of the whole trip, he made in the last two days of his travel. How many kilometers did Ron travel last summer?

9. **The length of a rectangular swimming pool is 24 meters.** A sidewalk, 1.5 meters wide is built around the swimming pool. Find the width of the swimming pool, if the area of the sidewalk around the swimming pool is 132 m².

10. **The sum of twice the smaller of two consecutive even numbers** and three times the larger consecutive even number is less than 188. However, the sum of three times the smaller number and twice the larger number is more than 174. What are these two consecutive even numbers?

11. **One out of 12 coins is fake, slightly lighter than the others.** How can you find which of these 12 coins is fake by using an old fashion "balance" scale without weights only 3 times?

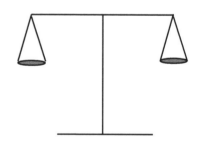

Level Three

Fun Home Assignment:

1. Each of the five students whose last names are Smith, Rogers, Berger, Sun, and Kim got awards for best project in one of the following subjects: Math, Science, English, French, and Art. The first names of these students are Alan, Maria, Anna, Ilya, and Foteh. Based on the clues below, please match each student's first name with their last name and with the subject for which the student got an award.

   - First name of Smith is neither Maria nor Ilya.

   - Smith congratulated Foteh for his excellent Art projects

   - Berger is not so good in Science

   - Alan does not like working on French projects

   - Anna tried to get better results than Maria in English and better results than Ilya in Math, but Maria and Ilya always received better marks than Anna in these subjects. Actually, Ilya was the best in Math and Maria was best in English.

   - Berger's first name is neither Maria nor Ilya

   - Kim did not get an award in English nor in Art

   - Foteh's last name is not Sun

   - Berger is not so good in Art

	Alan	Maria	Anna	Ilya	Foteh	Math	Science	English	French	Arts
Smith										
Rogers										
Berger										
Sun										
Kim										
Math										
Science										
English										
French										
Arts										

2. Translate from Base 10:

   a. 86_{10} = _____$_6$

   b. 86_{10} = _____$_4$

   c. 99_{10} = _____$_5$

   d. 143_{10} = _____$_{12}$

   e. 272_{10} = _____$_3$

   f. 647_{10} = _____$_9$

   g. 2319_{10} = _____$_7$

   h. 76_{10} = _____$_2$

Level Three

3. In △ABC, ∠A = 60°; side AB is split into 4 equal parts by points K, M, and L. The length of each part is 6cm. N is a mid-point of the side BC. NL is a perpendicular segment to AB. ∠LNM = 30°. Find perimeter of the △ABC.

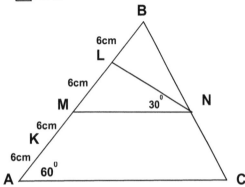

4. Peter and Dan are working on an assembly line for complicated toys. Dan is more experienced than Peter; he is assembling 7 toys at the same time as Peter assembles 4. There was an emergency order for 112 toys on Friday afternoon. The boss asked both Peter and Dan to come to work on Sunday to fulfill the order. On Sunday, Peter and Dan split the number of toys equally and they worked until all the toys were assembled. However, Dan assembled his 56 toys 2 hours before Peter finished his job. How many toys per hour do each of them assemble?

5. One out of 12 coins is fake but it is not known if the fake coin is lighter or heavier than the others. Can you find which of these 12 coins is fake using an old fashion "balance" scale without weights only 4 times? Can you determine if the fake coin is lighter or heavier than the other coins?

6. Solve the following "conjunction" sentences (assume X is an integer):

   a. $(2X - 5 < 0) \wedge (3X - 4.5 > 0)$

   b. $(7 - 5x \leq 27) \wedge (4X - 27 < (-3))$

7. Getting paid handsomely for their Sunday work, Peter and Dan stopped at the bar. Peter, being slower at work, was a more experienced beer drinker than Dan. For every 3 bottles Dan had, Peter managed to drink 5 bottles. They spent a considerable time drinking beer there. Peter finished his 17 bottles in 2 hours and 12 minutes before Dan finished his 17 bottles of beer. How many hours it took each of them to consume 17 bottles of beer?

8. The sum of four times the smaller of two consecutive odd numbers and the larger consecutive odd number is greater than 134. The difference between five times the larger of these two numbers and three times the smaller number is less than 66. What are these two odd consecutive numbers?

Level Three

9. Many Brain Power students are taking special exams to be accepted into prestigious schools. There are 25 questions in the Math part of the exam. For every correct answer students get 1 point (+1), for every wrong answer they are deducted 0.25 points (-0.25), and for every problem students did not attempt to solve there are 0 points given. This year Alex's mark was 15.75 points. How many problems did Alex attempt to solve?

10. Triangle ABC is isosceles; AB = BC. Base AC = 7.4 meters and height BD = 6.5 meters. MN is a mid-line connecting mid-points of the equal sides of △ ABC. An architect uses this as a model of the future tower she is proposing to build in downtown. To make the shape of the future tower unusual, the architect cuts the top of the △ ABC along the mid-line and rotates the top $60°$ to the right. What will the length of the dotted line connecting points B and D be?

11. Solve equations:

a. $|X-3| = |X-2|$

b. $|X+5| = |X+8|$

c. $|2X-1| = 2 \cdot |X-1|$

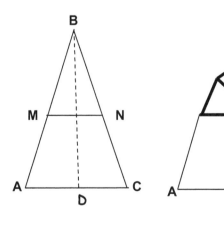

Level Three

LESSON 12

Classwork:

1. Translate from one Base to another:

   a. 154_{10} = _____ $_3$

   b. 375_8 = _____ $_{10}$

   c. 333_{10} = _____ $_4$

   d. 46_{10} = _____ $_2$

   e. 100010101101_2 = _____ $_{10}$

   f. $2C4E_{16}$ = _____ $_{10}$

2. Solve the following Exclusive "OR" riddle:

   This weekend, Simon and Jenny are celebrating their 40th wedding anniversary in their cottage on the beautiful lake. They invited their grown up kids Sandra and Victor for a small family event. At the sunset, they sat at the table on the beach and enjoyed nice food and some drinks. On the table there was a bottle of Scotch whiskey, a bottle of rum, red wine, and white wine. Each of the family members consumed only one kind of drink. Find what drink each of them had if:
   - Whiskey and rum were consumed by two men OR by two women
   - Sandra did not drink red wine
   - White wine and rum were consumed by Simon and Jenny OR by Simon and Victor
   - Simon's wife did not drink whiskey and she does not like red wine

   (Assume all four statements above are True)

3. In the trapezoid ABCD: BC || AD, BC = 17 cm, AD = 23 cm, AM = MB, CN = ND, BQ = MQ, CP = PN.
   Find the length of AR if QR || CD.

   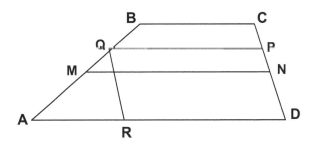

4. **Lenny is an auto mechanic and he works in the garage.** At the end of his working day he had to change oil in a small car that requires 4 litres of oil. That was when Lenny found out that he has only one jar of oil left that contains exactly 20 litres of oil. Lenny could not find an empty container that can hold exactly 4 litres of oil, but he has two empty containers, one can hold 3 litres and another can hold 5 litres. How can Lenny measure exactly 4 litres of oil using these containers so he would not waste the oil?

Level Three

5. **A circle is drawn in the trapezoid ABCD** so it touches sides BC, CD, AD, and segment BK. The radius of this circle equals 10 meters. It is given that AM = MB; MN || BC; MN = 6 meters, CD ⊥ AD, and BK ⊥ AD. Find the length of the sides AD, BC, and CD.

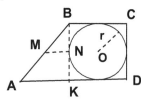

6. **A passenger ship was caught in a violent storm** and sunk near an inhabited island. However, many passengers managed to swim to the shore. The next morning they started looking for food and they found a few large coconut trees there. Since there were not too many coconuts on the trees, they decided that day to share the coconuts – one coconut for 3 people. The next day they decided to share each coconut among 5 people. On the third day they shared one coconut among 7 people. At the end of the third day no coconuts remained on the trees. How many passengers swam to the shore of this island if they consumed 142 coconuts in total in these three days?

7. **The Brain Power teacher experimented** in the class to see how fast her students can solve riddles with stickers. The teacher prepared 5 stickers: 3 white and two black. Then she called 3 students, Eden, Elina, and Elaine, to participate in the experiment.

a. First, the teacher blindfolded these students and she put black stickers on Eden's and Elina's foreheads and a white sticker on Elaine's forehead. The teacher asked students to raise their hands if they will find out the colour of the sticker on their foreheads after their blindfolds will be removed. The students are not allowed to talk to each other or to make any signs. They can only raise their hand.

After the teacher removed the blindfolds, the girls looked at each other and Elaine raised her hand immediately. A few seconds after, Eden and Elina raised their hands as well. All of them knew exactly the colour of their stickers. **Can you explain how the girls came to the right conclusions in this experiment?**

b. **In the second experiment**, the teacher attached a black sticker to Eden's forehead and white stickers to Elina's and Elaine's foreheads. After the blindfolds were removed, it took only a few seconds for Elina and Elaine to raise their hands. It took Eden another minute after Elina and Elaine raised their hands, to raise her hand as well. **Can you explain how the girls came to the right conclusions in this experiment?**

Level Three

c. **In the third experiment**, all three girls' stickers were white. They looked at each other for more than a minute, and then all three raised their hands simultaneously. **Can you explain how the girls came to the right conclusions in this experiment?**

The teacher was very impressed by the girl's ability to reason!!

8. **Two girls are comparing their earnings** for their work at Mr. Smith's farm last summer. Maria's daily pay was $32 and Anna's daily pay was $38. Anna worked 14 days less than Maria last summer, so her total pay was $262 less than Maria's.
How much money did each girl earn last summer?

9. **The school's auditorium is prepared** for a regional Math competition. For the first day of the competition, the janitor put 324 desks in equal rows (one student per desk). The next day, the janitor received an order to accommodate 111 more students. To meet this demand, the janitor added 3 more desks to each row and then he added two more rows of desks.
How many rows of desks were required for the first day of the Math competition if the number of rows on the second day was one less than twice the number of desks in a row?

(Use the formula for solving quadratic equations if needed:

$$x = \frac{-b \pm \sqrt{b^2 - 4ac}}{2a}$$)

10. **The manager of a marketing department** agreed to have a picnic for the employees in her department. The organizers bought enough food, so each person could get 4 hamburgers. However, 5 people did not come to the picnic. As a result of that, the participants had 5 hamburgers each. How many people came to the picnic?

11. **Calculate**:

a. 25034_6 43120_5
     $+\ 15542_6$ $-\ 24121_5$

b. $8B5D_{16}$ $B12A_{12}$
     $+\ 9CA5_{16}$ $-\ 75AB_{12}$

Level Three

Fun Home Assignment:

1. Five friends, Allan, Ben, Chris, Don, and Earl decided to spend their winter vacations in Florida. Since all of them live in different cities, Toronto, Ottawa, London, Calgary, and Montreal, one of the friends made reservations for hotel and car rental for all of them. The last names of the friends are Atkin, Brown, Cowen, Dimant, and Einman. Each of the friends have different hobbies: Math, Music, Sports, Movies, and Reading. Based on the clues below, please figure out each friend's full name, hobby, and the city he is from:
   - Einman lives in Montreal and he likes sports
   - Atkin and Cowen are not so good in Math and they do not like reading
   - The first letters of the friends' last names are in alphabetical order. The first letter of Don's last name is before the first letter of Allan's last name.
   - Don's and Ben's last names do not start with the letter "A"
   - Dimant does not like reading
   - The first letter of Ben's last name is before the first letters of the last names of Allan and Earl in the alphabet
   - The first letters of each friend's last name and first name are different
   - The friend from London likes reading
   - Atkin lives in Toronto and he likes movies
   - The first letter of the last name of the friend from Calgary is in the alphabet before the first letter of the last name of the friend from Ottawa.
   - The first letter of Dan's last name is before letter "C" in the alphabet.

   (Note: construct an appropriate logic chart to solve this problem)

2. **Translate from one Base to another:**

   a. 82091_{12} = _____ $_{10}$

   b. 235_{10} = _____ $_{2}$

   c. 320013_{4} = _____ $_{10}$

   d. 67514_{10} = _____ $_{8}$

   e. 10634_{7} = _____ $_{10}$

   f. 6287_{10} = _____ $_{6}$

3. **Calculate:**

   a. 1110110001_{2}
       $- 1001001011_{2}$

   b. 301121_{4}
       $+ 212233_{4}$

   c. $59A23_{11}$
       $- 3AA86_{11}$

Level Three

4. **Three married couples are sitting around a table.** The men names are Bob, Brian, and Ben; the women names are Sarah, Sandra, and Sofia. Can you figure out who is married to whom if the two following sentences are true:

   a. Ben is husband of Sarah and Brian is husband of Sofia, or Ben is husband of Sofia and Bob is husband of Sarah.
   b. Bob is husband of Sarah and Ben is husband of Sofia, or Bob is husband of Sandra and Brian is husband of Sarah.

5. **In the trapezoid ABCD:** BC || AD, CD ⊥ AD, AB = BD = 5m, CD = 3m. What is the length of the mid-line MN?

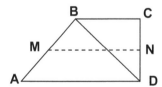

6. **Prove that diagonals in the isosceles trapezoid are equal in length.**

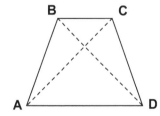

   Given: BC || AD, AB = CD.
   Prove: AC = BD

7. **A large music band of 198 musicians** entered a stadium in nicely formed columns with an equal number of musicians in each row. A few minutes later a second music band of 207 musicians entered the stadium and joined the first band. The two bands regrouped into one very large band having 5 more rows and 6 more columns than the first band had before the second band joined them. How many rows of musicians were in the first music band?

8. **The factory management is not happy** with the quality of a metal device they manufacture for a big client. The part consists of a head and a placeholder part for this head. Out of the batch that was inspected yesterday, 25 devices were defected. Some devices had defected heads only, and other devices had only defected placeholders. However, no device had both parts defected at the same time. In all, there were 17 good heads and 26 good placeholders.
   How many devices were in the batch that was inspected yesterday?

9. **It takes 12 fishermen 7 hours to catch 60 fish.**
   a. How many fish will 7 fishermen catch in 3 hours?
   b. How much time will it take for 9 fishermen to catch 45 fish?
   c. How many fishermen will catch 50 fish in 14 hours?

Level Three

10. Fill in the missing numbers:

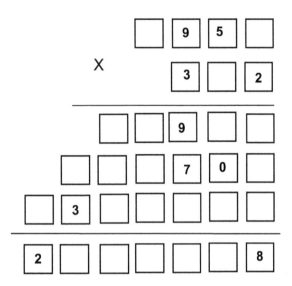

11. **Diagonal BD in the isosceles trapezoid** ABCD bisects angle ABC. Given that BC = 7 meters and the perimeter of the trapezoid ABCD equals 52 meters. Find the length of the mid-line MN.

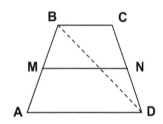

12. **When 7 is added to twice** the difference between a number and 5, the result is the same as if 3 is subtracted from twice the number. What is the number?

Level Three

LESSON 13

Classwork:

1. **The ten Brain Power students** that survived last year's encounter with witches, got together on the first anniversary of that event. (Those of you who do not remember, please see page 123 in the "Brain Power Enrichment: Level Two, Book Two"). Thanks to a clever plan and luck, all ten students survived that terrifying encounter. After a first toast for a lengthy life and for their parents that insisted their children must learn thinking skills, one of the students stood up and asked a question: "We were lucky to be a group of ten people in that crazy situation. The fact that we were ten people helped us find a clever solution for our survival. Would we be as lucky to survive if we were a group of 11 people? Is there a similar or compatible strategy for 11 people?" **What do you think? What strategy should help 11 people minimize their losses in the situation similar to the one the 10 students were in last year?**

2. Calculate:

   a. $D9ACF7_{16}$
          $- \underline{A8E239_{16}}$

   b. 625345_8
          $+ \underline{357623_8}$

   c. $9C8BD_{14}$
          $- \underline{7A3C2_{14}}$

   d. 111010011101_2
          $+ \underline{10001001111_2}$

3. **The tens digit of a two digit number is one less** than three times its unit digit. When tens and unit digits are reversed, four times the reversed number is 14 less than twice the original number.
   What is the original number?

4. **The unit digit of a two digit number is five less than the tens digit.** When tens and unit digits of this two digit number are reversed, twice the original number is 6 less than eight times the reversed number. What is the quotient of the original number divided by the reversed number?

5. Prove:

   $A \underline{\vee} B = (\overline{A} \wedge B) \underline{\vee} (A \wedge \overline{B})$

6. **The unit digit in a three digit number** is one less than the tens digit. The hundreds digit in this number equals to the sum of the unit and the tens digits. The number resulted from switching the hundreds

Level Three

digit with the unit digit is 297 less than the original number. **Find the original number.**

7. **In the trapezoid ABCD**: AB = CD, CK ⊥ AD, AK = 42cm, KD = 8cm, AM = MB, and CN = ND.
Find the length of MN.

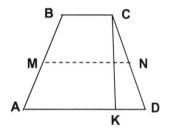

8. **Dave, a grade 2 student,** spent his summer vacations at his grandfather's farm. To improve his math skills, every morning Dave counted animals in his grandfather's backyard and their legs.
Yesterday he counted 17 heads and 46 legs in total. There were goats, dogs, and geese in the backyard.
a. **How many geese were in the backyard yesterday?**

This morning, Dave counted 32 heads and 72 feet in total. There were geese, chicken, dogs, but no pigs in the backyard.
b. **How many dogs were in the backyard this morning?**

Today, in the evening, Dave counted animals in the backyard again, and their legs. There were only pigs and chickens in the backyard. This time he counted 14 heads and 58 legs.

c. **How many pigs were in the backyard this evening?**

9. **In the right △ABC**, AC ⊥ BC, BC = 9cm, AC = 12cm, CD ⊥ AB. Find the area of △ADC.

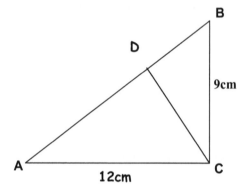

10. **Pavel has an equal number of nickels**, dimes, and quarters. How many coins does Pavel have in all, if the total value of his coins is $5.20?

11. **A small local theater** has an even number of seats. In the first twelve days of a Musical Comedy they sold more than 1700 tickets. In the next seven days they sold less than 1004 tickets. How many seats are in the theater if all the seats for every show were sold?

Level Three

Fun Home Assignment:

1. **a. Create multiplication tables** from 1 to 12_4 in base 4. Then answer:
   - $3 \times 3 =$ ___ $_4$
   - $10_4 \times 3 =$ ___ $_4$
   - $12_4 \times 2 =$ ___ $_4$

   b. Create multiplication tables from 1 to 13_7 in base 7. Then answer:
   - $6 \times 6 =$ ___ $_7$
   - $11_7 \times 6 =$ ___ $_7$
   - $13_7 \times 5 =$ ___ $_7$
   - $12_7 \times 4 =$ ___ $_7$

2. **Five students, Paula, Maria, Kathy, Dan, and Ben, wrote exams** in Math and English. Based on the clues below, please find out each student's marks on these exams:
   - The students' marks on both exams were 95, 90, 85, 80, 75
   - The student that received 90 in Math received 75 in English
   - Kathy's mark in Math was 10 points higher than Maria's mark in Math
   - Dan's mark in Math was 15 points less than his mark in English
   - Kathy's mark in English was 10 points less than Dan's mark in English
   - Paula's mark in English was higher than Ben's mark in Math

3. **Calculate:**

   a. 34241_5
      $- 13434_5$

   b. 21_5
      $\times 13_5$

   c. 8736_9
      $- 5867_9$

   d. $9B3C8_{14}$
      $+ 5A967_{14}$

   e. 213_4
      $\times 23_4$

   Calculate and then check your answer using Base 10.

4. **The tens digit of a two digit number** is two more than the unit digit. When tens and the unit digit are reversed, the sum of four times the original number and five times the reversed number is equal to 585. What is the sum of the digits in the original number?

5. **The difference between the tens digit** and the unit digit in a two digit number is 5. The difference between the original two digit number and twice the reversed number is 7. Find the reversed number.

Level Three

6. **Segment AB is a tangent to the circle** with the centre in point O. PR is the diameter of the circle. Distance from point P to segment AB equals 8.4cm and distance from point R to segment AB equals 14.6cm. What is the diameter of this circle?

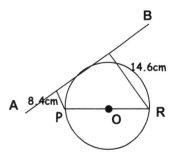

7. **In a four digit number**, the thousands digit is the same as the tens digit, and the hundreds digit is the same as the unit digit. When the unit digit is removed and is placed in front of the thousands digit, the difference between twice the original number and four times the new (reversed) number is 606 less than the reversed number. What is the original number if the tens digit in the original number is 4 more than the unit digit?

8. **In the isosceles trapezoid ABCD**, AB = BC = CD, ∠ACD = 90°. Find the length of the mid-line MN if BC = 12 meters.

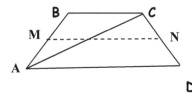

9. **John was a farmhand at Mr. McDonald's farm.** He signed a contract for 30 days work, for which he will get $120 and a bicycle. However, he had to quit after 20 days as he received an urgent phone call from his family requesting him to come to his niece's wedding. To keep up with John's contract, Mr. McDonald gave him a bicycle and $65. What was the bicycle's price?

10. **Four friends, Mike, Ben, George, and Leo** collected money for poor kids in Africa. Today they delivered their combined cheque to an agency that distributes donations to the kids in Africa who need meals and clothes to go to school. Mike collected half of the amount collected by his three friends, Ben collected $\frac{1}{3}$ of the amount collected by his three friends, George collected $\frac{1}{4}$ of the amount collected by his three friends, and Leo collected $260. What was the amount of the cheque the four friends delivered to the agency? How much money did each of the four friends collect?

11. **Two books and three pens cost $22.** Three books and two pens cost $23. How much does one book cost? How much does one pen cost?

LESSON 14

Classwork:

1. Five women, Anna, Betty, Cathy, Diana, and Emily live in different houses on the same side of a street. The house numbers they live in are 1, 3, 5, 7, and 9. Their husbands' names are Alex, Ben, Chris, Don, and Ed (of course, each woman is married to one man only). The women's professions are a nurse, a teacher, a business manager, an accountant, and a police officer. From the clues below, find out in which house each woman lives, her profession, and her husband's name.

   • Accountant and her husband Don live in house number 5

   • Don is not Emily's husband

   • Anna and Diana do not live in house number 1. Their husband's names are not Alex and not Don

   • Ben married the teacher and Ed is married to the business manager

   • Alex lives in the house number 3 and Chris's wife is the police officer

   • Diana does not live in house number 7. She is not a teacher nor a police officer

   • Emily does not live in house number 1

   • Betty is not an accountant and she is not Chris's wife

2. **The unit digit of a two digit number** is five less than three times the tens digit. When the tens and the unit digit are reversed, the reversed number is 20 less than twice the original number. What is the original number?

3. **The digits in a three digit number** are consecutive numbers with the hundreds digit being the largest among these three digits and the unit digit being the smallest. When unit and hundreds digits are replaced with each other, three times the reversed number is 60 more than twice the original number. What is the original number?

4. **In** △ ABC, AN and BM are medians, PN = 3m, PM = 5m, AB = 11m, and AC = 8m. Find perimeter of a. △ABP and b. △APM.

   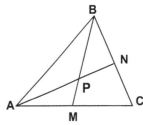

5. **In a Fibonacci sequence**, the fifth term equals 21. All the terms in this sequence are positive whole numbers.
   a. What is the sum of the first six terms of this sequence?

Level Three

b. What are the first two terms of this sequence?

6. **Triangle ABC is isosceles**: AB = BC = 12cm. Medians AN and CM are 9 cm each. Prove that △PNC is isosceles and calculate its perimeter.

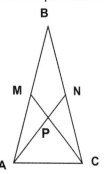

7. **An equilateral octagon's side** is 1 centimeter longer than the side of an equilateral triangle. The perimeter of this octagon is three times the perimeter of this triangle. Find the lengths of each side of the octagon and the triangle.

8. **Fira and Bohdan are working on a contract** at a large factory. Fira is paid $15 per hour and Bohdan is paid $13 per hour. Yesterday, Fira worked 3 hours more than Bohdan and she received $55 more than Bohdan for that day work.
How much money did Fira get for yesterday's work?

9. **Translate from base 10:**

   a. 1623_{10} = _____ $_8$

   b. 789_{10} = _____ $_4$

   c. 3782_{10} = _____ $_{12}$

10. **Maria had $6 more than Selina.** On Selina's birthday Maria gave $5 of her money to Selina. After giving the money gift to Selina, Maria's amount became half of Selina's new amount. How much money do both girls have altogether?

11. **The tens digit in a two digit number** is 2 more than three times the unit digit. Three times the reversed number is 2 more than the original number. What is the sum of these two numbers (original and reversed)?

Level Three

Fun Home Assignment:

1. **The tens digit in a two digit number** is larger than the units digit by 4. Nine times the sum of the digits is 10 more than the number. What is the number?

2. **The numerator of a fraction is 3 less** than the denominator. When the numerator is increased by 5 and the denominator is reduced by 5, the result is $4\frac{1}{2}$. What is the original fraction?

3. **The sum of the digits in a two digit number** is 9. The difference between five times the reversed number and the original two digit number equals the sum of the digits. What is the original number?

4. **The unit digit in a two digit number** is 2 less than the tens digit. The difference between the original number and the reversed number is 6 more than the sum of the digits in the original number. What is the reversed number?

5. **The difference between the tens digit** and the unit digit in a two digit number is 2. If the digits are reversed, the new reversed number is 4 less than 5 times the sum of the digits in the original number. What is the difference between the original and reversed numbers?

6. **The lengths of two bases BC and AD** in trapezoid ABCD are in ratio of 1 : 3. The length of the mid-line MN of this trapezoid is 18 meters. The shorter base is the side of equilateral triangle BCK built inside of this trapezoid. Point K is on side AD. What is the height and the area of this trapezoid (round to the nearest hundredth)?

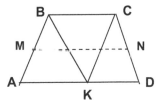

7. **Two families, Smith and Brown,** are preparing to celebrate a New Year party together. Both families agreed to buy wine and cakes for this party. The Smith family bought 3 bottles of wine and 2 cakes for a price of $33. The Brown family bought 2 bottles of the same wine and 3 cakes, same as the Smith family, for $32. Can you find out without using equations how much a bottle of wine costs and how much one cake cost?

8. Five old friends, Gleb, Lev, Alex, Sandra, and Victoria, finally decided to meet in a nice restaurant for a reunion party 20 years after finishing high school. Although they agreed to get together at 7pm, none of them came on

Level Three

time. Actually, the first of them arrived at 7:05pm, and the rest arrived one by one in intervals of about 5 minutes. Each of them ordered different drinks: wine, whiskey, rum, soda, and tequila. As they started conversation, they discovered that their professions are a teacher, an accountant, an engineer, a physician, and a truck driver. Based on clues below, please figure out the order in which they arrived to the party, the profession of each friend, and the drink each of them ordered:

- Victoria came after Lev but before Sandra
- The engineer ordered wine and the teacher ordered whiskey
- Alex arrived before Gleb. Alex was greeted by the engineer, Sandra, and by the teacher, when he came to the table.
- The physician ordered soda
- Lev does not like wine
- The friend that arrived last to the party ordered rum
- The accountant did not arrive first
- The physician remembered that Alex bullied her and the accountant at school

9. **Four friends, Brenda, Carmella, Ken, and Tom are collecting baseball cards.** Brenda has as many cards as Carmella and Tom together. Ken has as many cards as Brenda and Carmella together. Twice the number of Ken's baseball cards is equal to three times the number of Tom's cards.

A. How many times is the number of Brenda's cards more than the number of Carmella's cards?
B. How many times is the number of Tom's cards more than the number of Carmella's cards?

10. **Calculate**:

a. $\quad B39C_{14}$
   $- \underline{A8B9_{14}}$

b. $\quad 43_5$
   $\times \underline{\ 4_5}$

c. $\quad 7ECA_{16}$
   $+ \underline{98B7_{16}}$

11. **Four students, Teo, Victor, June, and Kevin** won $400 in a lottery. They split this amount among themselves in four different amounts. However, not everyone was happy with the amount they received as they perceived that the split was not fair. So they asked their friend Eddie, who was very good in problem solving, to redistribute the money they won so each student will get the same amount. Eddie agreed to help them but demanded to be compensated for his "intellectual" effort. The four friends agreed to this condition and Eddie gave the following orders:

- Victor must give his $15 to Teo
- Kevin must return $\frac{2}{3}$ of his amount

Level Three

- June must triple her amount

After these manipulations, each friend ended with the same amount and Eddie took the rest to himself as compensation for his hard work.

a. How much money did Eddie get?
b. How much money did each friend get initially?

Level Three

Level Three

LESSON 15

Classwork:

1. **Three professors, Dr. Smart, Dr. Cohen, and Dr. Avanti are from different countries, USA, Canada, and Italy,** are attending an international conference on environmental issues. Their specialties are chemistry, biology, and demography. Also, three PhD students with the names Smart, Cohen, and Avanti, are attending this conference as well. These students are also from USA, Canada, and Italy.
   - PhD student Avanti is from an Italian university
   - Professor of demography is from a Canadian university
   - Dr. Smart plays bridge sometimes with the professor of chemistry
   - Professor of demography and a PhD student who is a known champion in chess in his country, live in the same city and are members of the same golf club
   - A PhD student with the same name as the professor of demography lives in USA
   - PhD student Cohen never played chess seriously

   What is each professor's special field of study?

2. Calculate without a calculator:

   a. $\sqrt{2049.3729}$ =

   b. $\sqrt{15276.96}$ =

   c. $\sqrt{46785.69}$ =

   d. $\sqrt{35910.25}$ =

   e. $\sqrt{8.0656}$ =

3. **The unit digit in a certain two digit number** is 2 more than twice the tens digit. Twice the reversed number is 20 more than four times the original two digit number. Find the original number.

4. **The original four digit number is formed by consecutive digits;** the thousands digit is the smallest and the unit digit is the largest. In the reversed number, the thousands digit is the largest and the unit digit is the smallest. The difference between twice the original number and the reversed number is 1480. Find the reversed number.

5. **The unit digit in a two digit number** is 3 less than the tens digit. The difference between the original two digit number and the reversed number is five more than twice the sum of the

Level Three

digits in the two digit number. Find the original number.

6. **Find such a number that:**

   - When you divide this number by 3, the remainder is 1
   - When you divide this number by 4, the remainder is 2
   - When you divide this number by 5, the remainder is 3
   - When you divide this number by 6, the remainder is 4

7. **Trapezoid ABCD is isosceles**: AB = CD = 4m, the larger base AD = 7.6m, and ∠BAD = 60°. Make a drawing for this problem and find the length of the mid-line of this trapezoid.

8. **Given △ ABC. M and N are the mid-points** of sides AC and BC. Prove that all three vertices of this triangle are equally distanced from the line drawn through points M and N.

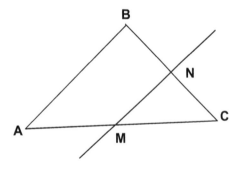

9. **Can you calculate the following?**

   a. 13_4
       $\times \underline{11_4}$

   b. 23_5
       $\times \underline{12_5}$

10. **A group of mountain climbers set a camp** near the top of the mountain. There is a steep road from a village at the bottom of the mountain to the camp but it is used very seldom, mainly to bring meteorologists to the top of the mountain. One day the group of mountain climbers sent a message to the village requesting a delivery of food and warm blankets. It took 7 hours for a truck with all requested supplies to drive up the mountain to the climbers' camp because the driver drove at a low average speed uphill. However, on the way back, the average speed of the truck was increased by 12 km/h and the truck returned to the village in 4 hours. **What is the distance from the village to the mountain climbers' camp?**

Level Three

Fun Home Assignment:

1. Four star musical groups, Classic, New Rock, Romantic, and Nostalgic Blues, are scheduling their concerts from April till September in four cities, Vancouver, Calgary, Ottawa and Montreal. Each group will give two concerts during the season in the same city. Based on the clues below, please figure out in which city each group will give concerts and during which months:
   - Only one group's schedule includes concerts in April and September. No other group schedule includes April or September.
   - There will be no concerts in Montreal in July
   - The first concert in Calgary will be in June
   - Nostalgic Blues are not giving a Concert in April
   - The Classic's two concerts are scheduled two months apart. None of their concerts will be in June and will not be in Calgary
   - The second concert in Montreal will be later than the New Rock's second concert in Ottawa
   - Only one group scheduled concerts for May and July. No other group schedule includes May or July.
   - There will be no concert in Montreal in August

2. Calculate without a calculator:

   a. $\sqrt{54.9081} =$

   b. $\sqrt{470.89} =$

   c. $\sqrt{64617.64} =$

   d. $\sqrt{31719.61} =$

3. **The sum of the digits in a two digit number is 12.** When the digits are reversed, the sum of 3 times the original number and twice the reversed number equals 312. Find the original number.

4. **The tens digit in a two digit number** is two more than the unit digit. The difference between the original two digit number and the reversed number is two more than twice the sum of the digits in the original number. Find the reversed number.

5. **Two planes started flying towards each** other at 10:15am from Toronto and from Montreal. The distance between Toronto and Montreal is approximately 600 kilometers. The average speed of the plane from Toronto is 105 km/h and the average speed of the plane from Montreal is 145 km/h. How far from Toronto will the planes meet in the air and at what time?

6. **Two trucks loaded with merchandise** started from Toronto at the same time. One truck is moving west by highway 401 to Windsor, Ontario and the other

truck is moving east by highway 401 to Montreal at a speed 15 km/h faster than the other truck. After 4 hours and 12 minutes, the distance between these two trucks is 525 kilometers. Find the average speeds of each truck.

7. Calculate:

   a. 26_7
      $\times \underline{14_7}$

   b. 35_6
      $\times \underline{23_6}$

   c. 46_8
      $\times \underline{32_8}$

   d. $B8_{12}$
      $\times \underline{A2_{12}}$

8. **Given isosceles trapezoid ABCD:** AB = CD = 20m, BC = 15m, and $\angle C - \angle A = 60°$. Find the length of midline MN of the trapezoid ABCD.

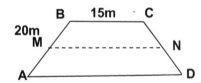

9. **A lion killed a zebra at 10:30am but decided to have a nap before eating the zebra.** When the lion just closed his eyes for a nap, a hyena passed by and stopped 10 meters away from the zebra realizing there was an opportunity for a free lunch. Being afraid of lions, the hyena would crawl forward only when the lion's eyes were closed. Each time the lion fell asleep, the hyena would advance 3 meters toward the killed zebra. The hyena would retreat 2 meters each time the lion opened its eyes. Assuming that the lion repeatedly closes its eyes for half an hour (having nap) and opens them for half an hour, at what time will the hyena have an opportunity to steal a piece of zebra?

10. **The unit digit in a four digit number** is one more than the tens digit. The hundreds digit of this four digit number is twice the tens digit. The thousands digit is one more than the unit digit. When the thousands digit and the unit digit are switched, the difference between the original number and the new "reversed" number will be 55.5 times larger than the sum of the digits in the original number. Find the original number.

11. **The length of a midline in a trapezoid** is 15 meters. The ratio of the bases in this trapezoid is 1 : 4. Find the length of each base in this trapezoid.

Level Three

LESSON 16

Classwork:

1. **The Grade 9 students of the Blue Heights high school** are preparing for this years' academic and art competitions. Yesterday they wrote tests in Math and English. Five students, Gleb, Alex, Nicole, Ansar, and Ofek are very excited because their marks on both tests are above 90%. Their marks in Math were 95%, 96%, 97%, 98%, and 99%. Their marks in English were 96%, 97%, 98%, 99%, and 100%. Based on the clues below, please find out what mark each of these students received on each of these tests.
   - Ansar's mark on the Math test was 2% higher than Ofek's mark on the English test
   - A student, who's mark on the English test was 99%, received 2% lower on the Math test than Gleb
   - The combined total of Alex's two marks was greater than 197
   - A student, who's mark on the Math test was 96%, received 2% lower on the English test than Nicole
   - Alex's mark on the Math test was different from Gleb's mark on the English test

2. **A truck driver is delivering construction materials** from the factory to the construction site, 566 kilometers away from the factory. It takes 6.3 hours for the whole trip. The truck driver drove at 82 km/h during the first part of the trip and at 95 km/h during the remaining of the trip. How long did he drive at 82 km/h?

3. **A military helicopter flew from the base** to the target at a speed of 110 km/h and immediately returned to the base on the same route at a speed of 78 km/h. How far from the base was the target if it took the helicopter 6 hours and 16 minutes for the whole trip?

4. **A bank robbery happened at 11:30am** at a branch near a highway. The robbers managed to escape in a waiting car and are driving at a speed of 110 km/h on a highway. Police were notified about the robbery only at 12:15pm. After talking to the branch staff at the branch, the police started chasing the robbers at 12:30pm on the highway at a speed of 135km/h. At 1pm the robbers decided to drive faster, at a speed of 120 km/h. **At what time will the police catch up with the robbers and how far from the bank's branch will this happen?**

5. **The unit digit of a two digit number** is 1 less than the tens digit. The difference between this two digit number and six times the sum of its digits is equal to the unit digit of this

two digit number. **What is the reversed number of this two digit number?**

6. **The date February 2, 2022 will be written as 2.2.22.** Another example is January 1, 2011. It will be written as 1.1.11. **How many times between year 2000 and 2100 will the date be written using the same digits?**

7. **Points A and B are located on the opposite** sides of segment PT. Distance from point A to the segment PT is 8 meters, and the distance from point B to the segment PT is 18 meters. Point K on segment AB splits AB into two equal segments. **Find the distance from point K to the segment PT.**

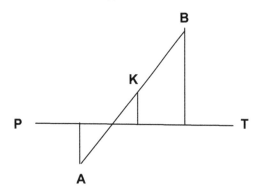

8. **Alex lives in Niceville and Dana lives in Besttown,** 420 kilometers away from Niceville. Last Sunday they decided to get together for a picnic. They called each other on the phone on Sunday morning and started driving immediately after their phone conversation. They arrived to the picnic stop at the same time, exactly 3 hours after talking on the phone. How far is the picnic stop from Alex's house, if Alex was driving 20 km/h faster than Dana?

9. **Martha has three boys** who are identical triplets. Their names are Jack, Jonathan, and Jerry. The boys are good students and they participate in various academic competitions. Last month they participated in Math and Essay Writing competitions at their school. As usual, they scored the top three places in those two events. Martha was very proud of her boys' achievements but she could not remember exactly who won what place. Please help Martha find out who won which place in each of the competitions if the following three facts are true:
   - Jack was third in Math and Jerry was the best in Essay Writing, or Jack was the best in Math and Jerry was third in Essay Writing
   - Jonathan was the best in Essay Writing and he was third in Math, or he was third in Essay Writing and the best in Math
   - Jerry was second in Essay Writing and Jack was the best in Math, or Jerry was second in Math and Jack was third in Math

10. **The digits in a three digit number** are consecutive even numbers; the hundreds digit is the smallest and the unit digit is the largest. The quotient of the number and the sum of its digits is 20.5. Find this three digit number.

Level Three

Fun Home Assignment:

1. **The Blue Heights high school** celebrates their students' achievement in competitions. Five students, Ken, Anna, Peter, Dennis, and Lindsay, competed in Bio (biology), Math, Music, French, and English regional competitions. Each of these students won first place in one of these subjects. Their marks were 93, 94, 95, 96, and 97. These students' last names are Kim, Levy, Meron, Gutman, and Levitan. Read the clues below and find out each student's last name, the subject in which this student won first place in, and the winning mark for this subject.
   - Meron was the best in the Bio competition
   - The highest mark was in the French competition
   - Levitan won the Music competition
   - The winning mark in Math was higher than the mark received by Lindsay
   - Anna did not win the Bio competition
   - Gutman did not win the French or the English competition
   - Peter's mark was lower than Ken's mark but it was 3 points higher than Anna's mark
   - Lindsay's mark was lower than Levitan's mark
   - The number of letters in the student's first name is different than the number of letters in the student's last name, than the number of letters in the subject this student won, and, than the last digit of this student's mark on the student's winning subject

2. **Two trucks started moving at the same time** from Toronto on highway 401; one truck was going to Montreal and another to Detroit. After 6 hours the trucks were 900 kilometers apart from each other. Find out the speed of each truck if the speed of the truck going to Montreal was 10 km/h more than the speed of the truck going to Detroit.

3. **Sophie lives in Toronto** but Sophie's daughter, Carolina, is a student in Montreal. Once a month Sophie drives to Montreal to visit Carolina and to bring her some food and books from home. Last Sunday, Sophie decided to drive to Montreal. She loaded her car with stuff for Carolina and started driving to Montreal at 8am. At 10am, Sophie received a phone call from her husband Matthew. Matthew informed Sophie that she forgot her credit cards at home and that he is driving to bring the cards to Sophie. He asked his wife to continue driving to Montreal at the same speed she was driving so far so he will be able to catch up with Sophie at 2pm. How far from Toronto will Matthew catch up with his wife if his speed will be 35 km/h more than Sophie's speed?

Level Three

4. **The length of median BD in △ABC** is the same as the length of AD and DC. AM is another median in this triangle. ∠ABD is twice larger than ∠DBC. Find the perimeters of △AOD and △ABC if BD = 12 meters. (Calculate to the nearest whole number).

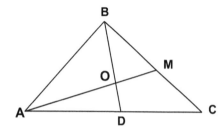

5. **A cyclist started riding at 8am** at a certain speed. He made his first stop after being 3.5 hours on the road. After resting for 18 minutes, the cyclist continued riding for another 2.2 hours (2 hours 12 minutes) at a speed of 4 km/h higher than in the morning. What was the cyclist's average speed during the first 3.5 hours and what was his speed during his last 2.2 hours if he drove 105.7 kilometers in total that day?

6. **The units digit of a two digit number** is 2 more than the tens digit. When the unit digit and the tens digit are reversed, the new number is 3 more than 6 times the sum of the digit in the original number. What is the sum of the digits in this number?

7. Students in SmartBrainyPower high school are exhibiting their moving robots. At the end of the exhibition they organized a running competition among the robots. Girls decided to race their robots against robots made by boys. The girls' robots names are Beautiful, Light, and Wise. The boys' robots names are Great, Mighty, and Scary. The boys' robots are higher than the girls' robots. Each pair of robots race twice against each other:

   a. In the first round, Beautiful run against Great. The Beautiful's step is 20% shorter than the Great's step but she makes 20% more steps per minute than Great. Who won this race?

   b. The girls tuned Beautiful before the second race. This time Beautiful makes 30% more steps per minute than Great. Who won this race?

   c. Light is racing against Mighty. Her step is 30% shorter than Mighty's step. In the first round, Light is making 30% more steps per minute than Mighty. Who won this race?

   d. The girls tuned Light for the second race against Mighty. Now Light is making 40% more steps per minute than Mighty. Who won this race?

   e. In the last competition Wise race against Scary. Her step is 10% shorter than Scary's but she makes

Level Three

10% more steps per minute than Scary. Who won this race?

f. For the next (last) race Wise was tuned to make 20% more steps per minute than Scary. Who won this race?

8. **Jason's mom stores all washed men socks** in a box in the laundry room. Yesterday there were 100 socks in all: 24 blue socks, 16 grey socks, 32 brown socks, and 28 black socks. Yesterday night the whole family was dressing for a party when suddenly all the lights went off due to a blackout. Mom asked Jason to bring 3 pairs of black socks for 3 men, dad, Jason, and Jason's brother. "How will I find exactly 3 pairs of black socks in such darkness? I will have to bring all the socks from the box!" exclaimed Jason. "No, Jason" said his mom. "Bring me a minimum number of socks to be sure you have at least 3 pairs of black socks. Now think, how many socks do you have to pull out of the box? You have enough Brain Power to figure this out."

a. How many socks will Jason pull out of the laundry box to be sure he brings at least 3 pairs of black socks?

b. How many socks will Jason pull out of the laundry box to be sure he brings at least 2 pairs of grey socks?

c. How many socks will Jason pull out of the laundry box to be sure he brings at least 4 pairs of brown socks?

d. How many socks will Jason pull out of the laundry box to be sure he brings at least 2 pairs of socks (each pair must be the same color but different pairs may be different colours?

9. **A taxi driver spent 4 hours driving people** from the hotel to the airport. He drove 316 kilometers in all in these 4 hours. In the first 1.5 hours, his speed was 8 km/h lower than in the next 2.5 hours. How many kilometers did the taxi drive in the last 2.5 hours?

10. **Among 80 coins there is one fake**, heavier than the other coins. You have an old fashion scale without weights. Your task is to find the fake coin among these 80 coins by using this scale only 4 times.

11. The arrow below is made up of 16 sticks. Move around 8 sticks to create 8 equal (congruent) triangles.

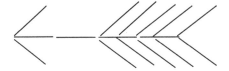

Level Three

Level Three

LESSON 17

Classwork:

1. **Three friends, Ariel, Ben, and Carmel,** travelled during last summer to six cities: Ottawa, London, Liverpool, Paris, Prague, and Moscow. Each friend visited two of these places. One of the friends was the only person who visited cities whose name starts with the same letter. Ariel visited Liverpool but not Moscow. Carmel visited Prague. No two friends visited the same city. Who visited which city last summer?
This problem has a definite solution!

2. Find the value of each angle indicated by a letter:

   a.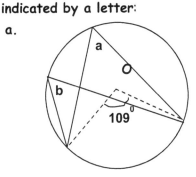

   Letter O indicates the centre of the circle.

   b.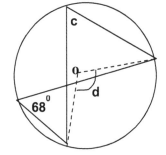

   c.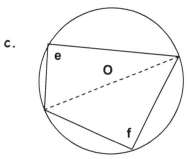

   d.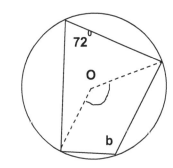

   e.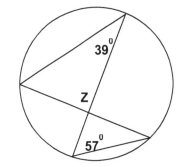

3. **A plane took off from Toronto airport** at 11am in direction to Vancouver. Another plane took off from Toronto at 12:30pm in direction to Halifax. By 2pm the distance between these two planes was 1965 kilometers. The velocity (speed) of the plane to Halifax was 130km/h less than the speed of the plane heading to Vancouver. What are the speeds of each plane?

Level Three

4. **Emily decided to visit her grandfather** who lives on a farm 557.6 kilometers from her house. She drove at 83 km/h for some time until she heard some strange noise in the car engine. At that moment Emily slowed down to 55km/h and continued at that speed until she arrived to her grandfather's farm. Emily was very tired as the whole trip took 8 hours. How long did Emily drive at the speed of 55km/h?

5. **Calculate by hand** (without calculators):

   a. $\sqrt{23808.49} =$

   b. $\sqrt{5416.96} =$

   c. $\sqrt{1204.09} =$

6. **Two fishing boats in a sea were 262.5** kilometers away from each other. The boats started moving toward each other and met after 7.5 hours. What were their speeds if one of these boats moved 9km/h slower than another?

7. **Calculate without using pens, pencils,** or calculators:

   a. $148^2 - 147^2 =$

   b. $236^2 - 235^2 =$

   c. $512^2 - 511^2 =$

8. **Michelle bought 5 books**, 2 pens, and 1 calculator on sale for the total amount of $32.50. Her friend Rose bought 3 books, 6 pens, and 7 calculators on the same sale for the total amount of $55.50. What will be the combined cost for three items: one book, one pen, and one calculator?

9. Peter drove 8 hours from his town to the city to see the major league hockey game. If Peter would drive 10km/h faster, he could get to the hockey game in 7 hours. At what speed did Peter drive to the hockey game?

10. **Christine is twice older now than Judith** was when Christine was as old as Judith is now. How old is each of them now if their combined age now is 35 years?

Fun Home Assignment:

1. Five friends consisting of two couples and a single man are discussing their travelling plans for next summer. The friends' names are Kathy, Richard, Dave, Teodora, and Kevin. They want to travel to France, Portugal, Sweden, Denmark, and Holland, however, each of the friends wants to travel to a different country.
   - The single man wants to travel to Holland
   - Kathy's husband wants to visit Portugal
   - Richard decided not to travel to France although his relative decided to go there this summer
   - Dave's wife wants to travel to Denmark
   - Teodora's husband wants to travel to Sweden
   - Kathy does not want to travel to Denmark
   - Kathy is not happy about her husband Kevin's choice of travel

   Who is the single man among these five friends? Who wants to travel to which country?

2. **Two little rabbits, Rick and Rock,** decided to leave their parents nest and explore the world. They agreed that Rick will go north while Rock will go south. Each took a cell phone to keep in touch. Being larger among the two brothers, Rick moved at a speed 3km/h faster than Rock. Eighteen hours after they left home, they called each other and they figured out that they are 486 kilometer apart from each other. What was each of the brother's average speed on their journey?

3. **The distance between city A and city B** is 774 kilometers. At 8am, a bus started from city A in direction to city B and at the same time a truck started from city B in direction to the city A at a speed of 14km/h more than the bus from city A. What was the speed of the bus if the bus and the truck met on the road at 12:30pm?

4. **At 7:30 am, Sarah started riding her** bike to visit her friend Dana near the foot of the mountain. Forty five minutes after Sarah's departure, her mom noticed that Sarah left her cellphone at home. Mom asked Sarah's brother Mike to take a bike and catch up with Sarah to give her the cell. Mike started immediately at a speed 4km/h higher than Sarah and he gave Sarah her cellphone at 9:45am. How far from their home did Mike catch up with Sarah?

5. **Two trucks started moving toward each other** at the same time from two different cities. One truck can arrive to its destination in 9 hours while another truck can arrive to its

Level Three

destination in 12 hours. After being on the road for 4 hours, the trucks are still 192 kilometers apart from each other.

What is the speed of each truck and what is the distance between these two cities?

6. **Find out the value of each angle or arc** indicated in the diagrams below:

a.

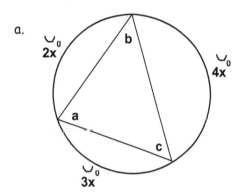

b.

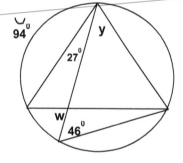

c.

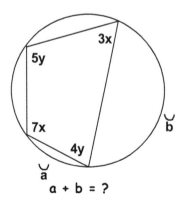

a + b = ?

7. **Calculate:**

a. $\sqrt{78064.36}$ =

b. $\sqrt{57.198969}$ =

c. $\sqrt{9.8695877}$ =

Calculate below without a pen, pencil, or calculator:

d. $87^2 - 86^2$ =

e. $242^2 - 241^2$ =

f. $56^2 - 54^2$ =

g. $27^2 - 24^2$ =

8. **Two trucks are delivering goods** from a factory to a distribution warehouse in another city. A young truck driver can complete a delivery in 8 hours, but the older truck driver can do the same in 6 hours. Knowing that he drives slow, the young driver started his journey at 7am. The older driver did not rush, so he started at 8am. At what time will the older driver catch up with the younger driver?

9. **Ron is in grade 5 and he likes doing** complex multiplications whenever there is an opportunity. He opened a heavy encyclopedia randomly and noticed two facing page numbers. He multiplied these numbers and

remembered the product, 6006. (It is easy to remember this palindrome number!) The next day Ron wanted to find the pages he opened a day before, but he did not remember the page numbers; he remembered only the product of these pages. Can you help Ron find out what the page numbers were?

10. **Four projects made by Grade 8** students are on display. These are the finalists in the latest competition for the most aesthetic model of a future condominium. These models are in the shape of a cube, pyramid, cone, and a cylinder and they are beside each other. A few of the models have lighting and a few are made of a recyclable material. Each of the models are painted either blue, green, or yellow. The first prize was given to a yellow model that was made of recyclable material and that had lighting. Please figure out which model won the first prize based on the clues below:

- The cube shaped model is made of recyclable material **OR** on either side of the pyramid shaped model are the only two models that have lighting.

- The only two blue models are standing on either side of the cone shaped model.

- The pyramid and cone models are painted the same colour and the cube shaped model is yellow, **OR** the cylinder and cone shaped models are the only models made of a recyclable material and the cube shaped model is painted green.

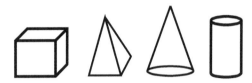

Projects on display

Level Three

Level Three

LESSON 18

Classwork:

1. Five students, David, Daniel, Amy, Sherry, and Rachel participate in a provincial exchange program during the summer time. These students live in different cities, Ottawa, Toronto, Kingston, Belleville, and London. As part of the exchange program, each of these five students spent last summer away from home in one of these cities. Please review the clues below and find out in what city each of these five students live in and in which city each of the five students spent last summer:
   - Amy did not spend summer in London
   - Rachel lives in Kingston
   - David spent the summer in Ottawa
   - Sherry does not live in Toronto nor in London and she did not spend the summer in these cities
   - Rachel spent the summer in the city in which Daniel lives in
   - Sherry lives in the city where Daniel spent his summer time

2. **The length of the chord AB is the same** as the length of the radius R.

   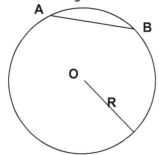

   Prove that the measure of the arc AB is $60°$.

3. **Diameter CD is perpendicular to the** chord AB. Find measures of the arc AD, and angles x and y.

   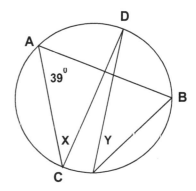

4. **AB is a tangent line to the circle.** AB touches the circle at point T. Length of the chord TD is half of the length of the diameter TC. Find the measures of ∠BTD, ∠X, ∠Y, and ∠Z.

   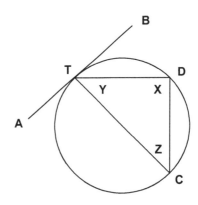

5. **Eddie can row a boat at an average speed** of 12km/h in still water (in a lake). Yesterday he rowed his boat on the river from home to another town,

Level Three

32 kilometers down the river, and then he returned home. It took Eddie 6 hours for the whole trip. What is the rate of the river?

6. **The speed of a plane in still air**, when there is no wind, is 145km/h. Yesterday it was very windy, so when this plane flew with the wind from city A to city B, it took 6 hours for the trip. On the way back, against the wind, the plane covered only $\frac{21}{37}$ th of the distance it made from city A to city B in 6 hours.
   a. **What was the speed of the wind yesterday?**
   b. **What is the distance from city A to**
      city B?

7. **This is an old classic problem** about a curious fly. Two cyclists living 15 kilometers away from each other agreed to get together for a picnic in a place exactly 7.5 kilometers from each cyclist's house. Assume that the cyclists' houses and the picnic place are on the same straight line. The cyclists started pedaling toward the picnic place at the same time, 8 am, and at the same speed, of 15 km/h. As the cyclists start moving, a fly sitting on one of the cyclist's shoulder became interested in what the other cyclist is doing. The fly took off at a speed of 30km/h toward the other cyclist. At the moment the fly touched the second cyclist's shoulder, it flew back to the first cyclist, and then back to the second, and on and on until the two cyclists arrived to the picnic area. What was the total distance the fly flew that morning?

8. **Peter, Frank, and Dave are preparing for camping in the mountains.** They agreed that Peter and Frank will take care of the food supply. Peter managed to buy 50 packs of frozen food and Frank bought 40 packs. Dave could not do shopping as he was busy preparing for exams. Instead, Dave gave his friends $90 for his share of food. What part of these $90 should Peter take and what part should Frank take if the food will be shared equally among all three friends?

9. **A clerk in the Lost and Found department** at the airport received a phone call from one disgruntled passenger. The passenger complained that all his luggage, two suitcases and three carry-ons did not show up on the conveyor at the airport. The clerk asked the passenger to provide some facts about the luggage, like the colour of each suitcase was and its weight. The passenger provided the following facts:
   • The colours of the five pieces of luggage are red, green, black, yellow, and grey. Each piece was a different colour.
   • The passenger remembered that the total weight of the luggage was 65 kilograms. However, he did not

Level Three

remember the exact weight of each piece.
- The passenger remembered that the difference in weights between the red and green pieces was the same as the difference between the weight of the black and yellow pieces.
- The weight of the green carry-on was one fifth of the weight of the red suit case.
- The weight of the green suit case was five times the weight of the yellow carry-on.
- Each piece of luggage's weight was in whole kilograms.

"I think that I will be able to figure out the weight of each of your piece of luggage using the facts you have provided. Could it be that one of your suitcases weighs 20 kilograms?" asked the clerk. "No" answered the passenger. OK" said the clerk. "Now, give me your address and phone number so we will be able to contact you when we find your baggage."
Can you find out the weight of each piece of luggage?

10. **Helen's son is studying at a university** away from home. Last week Helen decided to visit her son. She started driving from her home to the university where her son is studying, at 10am at a speed of 85 km/h. After spending 3 hours with her son, she returned home by the same road at a speed of 68km/h due to the bad weather. How far is the university from Helen's home if she returned home at 10pm that day?

Level Three

Fun Home Assignment:

1. **Thirty students from a grade 8 class** visited an apple orchard at Mr. Smith's farm. They were allowed to take some apples home. One of the students took 12 apples and other students took less than 12 apples each. Prove that there are at least 3 students in this grade 8 class that took home the same number of apples.

2. **Three runners, Nicholas, Genya, and Ilia**, were competing in a 24 kilometer race. When Nicholas crossed the finish line, Genya was 4 kilometers away from the finish line while Ilia was 8 kilometers away. How far from the finish line will Ilia be when Genya crosses the finish line?

3. **The length of the chord AB** is the same as the radius of the circle. Radius OT is perpendicular to AB. What is the length of the segment AT if the radius equals 24 centimeters?

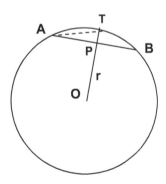

4. **Quadrilateral MTPN is inscribed** in the circle. AB is a tangent line touching the circle in point T. $\angle ATM$ equals $54°$, and $\angle TMN$ equals $82°$. Find values of $\angle PTB$ and $\angle MTP$ if side PT equals to side PN.

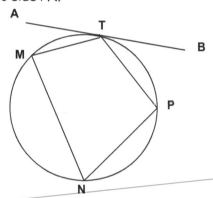

5. **Two friends live in cities 102 kilometers** apart. These cities are located on the same river. Both friends own motor boats. One boat's average speed in still water is 22km/h and the other boat's speed in still water is 20km/h. The friends decided to meet on the river somewhere between their cities. The friend, who has a faster boat, started riding his boat upstream at 9am and he arrived to the meeting place at 12pm. The other friend started moving at 10am downstream and arrived to the meeting place at 12pm as well. What is the rate (speed) of the river?

6. **A company's jet is flying to the company's** offices overseas. It took the jet 12 hours to get to the offices.

On the way back, the pilot increased the speed by 90km/h and the jet returned to the home base in 10 hours. What was the jet's speed to the offices? How far away the offices from the home base?

7. **One week after the trip** (see problem #6), the company's jet was taken to an island in the Caribbean and back on the same day. This time the jet was programmed to fly 405km/h in still weather (no wind). However, there was a constant wind that day. It took the jet 6.5 hours to get to the destination in the Caribbean, but 7 hours to get back to the home base. What was the speed (rate) of the wind that day? How far was the island from the home base?

8. **Each of the five friends, Max, Raz, Maya, Dana, and Dasha**, are working hard to prepare exhibits for their science project. Each exhibit consists of three parts: a stand, charts and formulas, and short essays explaining the research subject. On Friday, the friends stored parts of their work in a locker with the intent to assemble their exhibits next Monday afternoon for a Tuesday morning presentation. Unfortunately, they were busy all of Monday and could not work on their exhibits, so they asked their friend Jimmy to assemble their exhibits to be ready for Tuesday morning. To their surprise, on Tuesday morning they discovered that Jimmy mixed up the parts of their exhibits in such a way that each exhibit he assembled contained work of three different students. Based on the clues below, please find out how the five exhibits were assembled by Jimmy:
   - Maya's essay is not on Dana's stand
   - Raz's essay is with Dasha's charts
   - Raz's essay and Dasha's chart are not on Maya's stand
   - Max's essay is with Dana's charts
   - Maya's charts are on Dasha's stand
   - Max's essay and Dana's charts are not on Maya's stand

   (This problem is borrowed from the Level Two, Book Two)

9. **A truck is moving at a speed of 30km/h**. The driver wants to increase the speed of the truck to save one minute per each kilometer.
   a. By how much will the driver have to increase the speed of the truck?
   b. By how much will the driver have to increase the speed to save 2 minutes per each kilometer?

Level Three

10. **Given circle with radius 10cm.** AD is the diameter. PO $\perp$ chord AB and OS $\perp$ chord AC. PO = 7cm and OS = 5cm. As well, BD $\perp$ AB and DC $\perp$ AC.
   a. Find the perimeter of quadrilateral ABDC (round to whole number)
   b. Find value in degrees of the arc DC

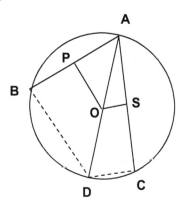

Level Three

LESSON 19

Classwork:

1. **John and his apprentice Alan won a contract** to build a barn on the farm. If John would build this barn alone, it would take him four times less days than Alan, if Alan would do the same job alone. Working together, they will build the barn in 28 days. How long would it take Alan alone to build the farm?

2. **A hungry squirrel, Nick said to his friend**, squirrel Rick, that he can eat all the nuts from his almond tree in 15 hours. Rick, to impress his friend Nick, responded that he can empty the tree in 12 hours. "Wow", exclaimed Nick, "Let's make a deal. Since you eat faster than I do, let me enjoy 2 hours eating alone, and then you will join me in the festivity." Rick looked at his watch; it was exactly 10am. He agreed to this proposal and Nick started eating nuts first. At what time will there be no more nuts on this tree?

3. **Michal ran an experiment in her bathtub.** She closed the drain at the bottom of the bathtub and opened the shower head. It took 18 minutes to fill the bathtub. Michal turned off the shower head and opened the drain at the bottom of the bathtub. It took 24 minutes to empty the bathtub. Michal started thinking: how long will it take to fill the bathtub if she will open the drain and then turn the shower head on?

4. Find values of x and y:

   a.
   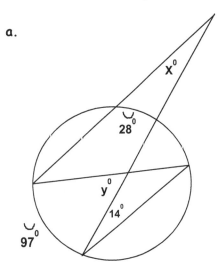

   b. <u>Given</u>: AT is a tangent segment.
   AT = 16cm, AB = 21cm,
   CN = 14cm
   Find length of AN (x) and AM (y)
   (round your answer to the tenth digit)

   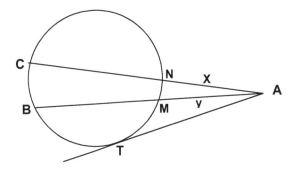

Level Three

5. **Four men, Ofek, Charles, Bernie, and Jeremy** met at the school reunion party. They were happy to see each other after so many years! While talking, they found out that each of them lives in a different city, Toronto, Calgary, Montreal, and Vancouver. Each of them has a different profession: a teacher, an engineer, a dentist, and a pilot. All of them are married and their wives names are Julia, Nicole, Maria, and Lisa. Based on the clues below, please find out the profession of each man, wife's name, and city he is from:
   - The teacher's wife's name is Maria
   - The engineer lives in Vancouver
   - Ofek does not live in Toronto
   - Jeremy is not a teacher
   - Lisa and her husband do not live in Toronto
   - Nicole and her husband live in Vancouver
   - Julia and her husband do not live in Montreal
   - Charles was very happy to meet his childhood friends from Toronto and from Vancouver
   - Nicole and Julia are not married to Bernie nor to Jeremy
   - Lisa's husband is a pilot

6. **A truck driver, Jonathan, was on-time** with the delivery of a container to the factory in another city. However, after driving half way to his destination, he increased his speed by 30% and he arrived 45 minutes ahead of the scheduled time. How much time did Jonathan spend on this trip?

7. **Michal and Karin are cleaning the floor** in the library. Michal alone can do this job in 1 hour and 30 minutes, but Karin can do the same job in 1 hour and 15 minutes. Yesterday they started the cleaning job together, but Michal received an emergency call after working for 15 minutes and she had to leave. How much time did it take Karin to finish the job? (Round to the nearest tenth)

8. **Spring arrived and Anna's family decided to prepare their cottage for the summer season.** They piled their car with cleaning materials and started driving at 9:00 am to the cottage 208 kilometers away from their home. At the beginning, they moved at 70 km/h because there was a traffic jam; there were too many cars moving in the same direction. Later they increased their speed to 95 km/h and they arrived at the cottage at 11:24 am. How much time were they driving at 95 km/h?

9. Find the missing digits:

```
      □ 4
   x  3 □
   _____
    1 □ □
    □ □
   _____
    8 □ 4
```

Level Three

Fun Home Assignment:

1. **Mike can paint a house in 12 hours.** His helper Dave can do the same job alone in 18 hours and their apprentice Jim can do this job alone in 24 hours. Mike started this painting job together with Jim, but after 2 hours they were called to do a painting job at another location. Mike asked Dave to come and finish painting the house alone. How much time will it take Dave to finish the job?

2. **Pipe A can fill a tank in 25 minutes.** Pipe B can fill the same tank in 20 minutes. To fill the empty tank, pipe A was opened alone for 3 minutes and, after that, pipe B was also opened. How much time will it take for both pipes A and B to finish filling the tank?

3. **Robot A can assemble 50 toys in 4 minutes.** Robot B can assemble 22 toys in 3 minutes, and robot C can assemble 45 toys in 6 minutes. The factory received an order for 2460 toys. How long will it take the three robots working together to complete the order?

4. **A bicyclist is riding his bike through a narrow tunnel at his maximum speed.** When he is $\frac{3}{5}$ through the tunnel, he noticed a huge truck is approaching the tunnel at a speed of 80km/h. The scared cyclist makes quick calculations: if he continues riding his bike at the same speed, he will be hit by the approaching truck exactly at the end of the tunnel. If he turns back and continues moving at the same speed, the truck will hit him from behind at the entrance to the tunnel. What was the cyclist's speed?

5. **Given external angle EAB.** AT is a tangent line; length of AT = 18 cm. $\angle P = 28°$; $\angle S = 74°$. Length of AB = 6 cm, and length of ED = 77 cm. Find out the value of $\angle$ EAB, and the length of EA and CA.

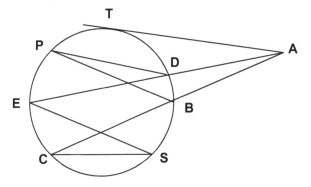

6. **Pipe A together with pipe B can fill a tank in 45 minutes.** Pipe B and pipe C can fill the same tank in 60 minutes. Pipe A and pipe C can fill the tank in 90 minutes. How long will it take the three pipes opened together to fill the tank?

Level Three

7. **Two trucks were delivering merchandise from city A to city B, 504 kilometers away.** They started moving at 9am. One truck's load was much heavier than another, so it moved at 4km/h slower than the lighter truck. During the trip, the truck driver of the faster truck rested for 1 hour but the truck driver of the slower truck rested only for 42 minutes. Both trucks arrived to the destination at city B at the same time. How long was their journey?

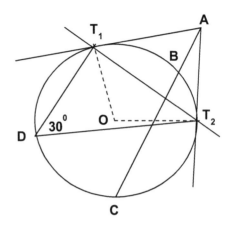

8. **Fill in the missing digits:**

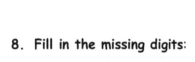

9. **Given point A outside of the circle.** AT_1 and AT_2 are tangent segments to the circle from point A. OT_1 and OT_2 are radiuses of the circle. Length of AB = 8cm and of BC = 24cm. Also given that AT_1 and AT_2 4cm longer than radius of this circle.
∠ T_1DT_2 = 30°.
Find: a. area of this circle
b. ∠ T_1AT_2 = ?
Prove: △ AT_1T_2 is isosceles

10. **Volume of water tank A is the same** as of tank B. Water tank A can be filled up by the pipe above it by 5 hours. Full tank A can be emptied in 8 hours when the pipe above it is closed but the pipe at the bottom is opened. Tank B can be filled up by the pipe above it in 4 hours. Full tank B can be emptied in 10 hours when pipe at the bottom is opened. Assume both tanks are empty and all the pipes above and below tanks A and B are open at the same time. What part of tank A will be filled up when tank B is $\frac{1}{3}$ full?

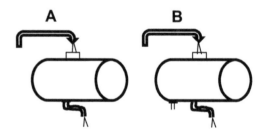

Level Three

LESSON 20

Classwork:

1. **A movie theater can hold 220 viewers.** A regular ticket's cost is $12 and a senior's ticket's cost is $7. Yesterday the theater was packed and all the tickets were sold. The total revenue from the sale of tickets was $2265. How many senior's tickets were sold yesterday?

2. **Elizabeth decided to invest her $8000** bonuses she received at the end of the year for excellent work in the company. She split this amount into two unequal parts. One part she invested in 6% simple interest and the other part in 5% simple interest. At the end of that investment year she received a total of $430 interest income from both investments.
How much money did Elizabeth invest at 6%?

3. Owner of a Natural Food store received an order for 9 kilograms of mixture of almonds and peanuts. The price for almonds is $16/kg and for peanuts is $7/kg. How much almonds and how much peanuts must the store owner mix to sell this mixture for $10/kg?

4. **How much filtered water must be added** to 750ml of 64% alcohol to reduce it to 40% alcohol?

5. **Chords AD and CB intersect in point E.** $\angle BAD = 32°$, and $\angle ADC = 44°$. CE = 7m, EB = 9m, and AE = 5m.
Find: a. length of AD
b. value of $\angle BED$

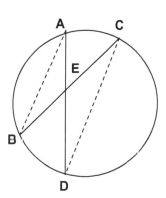

6. **Nicholas is late for his date with Michal.** He grabbed his bike and left his house 10 minutes before set time to meet Michal at the city square which is located 3 kilometers from Nicholas's house. First 1.5 kilometers Nicholas pedaled at a speed of 12km/h. At what speed will Nicholas have to pedal the rest of the distance to the place of his meeting with Michal to be exactly on time for their date?

7. **A farmer alone can plow his field in 4 days.** Together with his farmhand they can plow the same field in 3 days. In haw many days can the farmhand alone plow the same field?

111

8. **All three men in the Brown family**, father and his two sons, were painters. Last spring and summer they got a contract to paint three identical buildings. The first building was painted by the father alone in 45 days as his sons were still in school.

   The second house was painted in 27 days by the father together with his oldest son. The third house was painted in 36 days by the father together with his youngest son. In the fall there was a request to paint another house identical to those painted in summer. This time the father had to go away to a painters' conference, so he asked his two sons to do the job. How many days will be required for the two sons to paint such a house working together?

9. **Four married couples met for a lunch** at the International Food restaurant. The four men's' names are Ansar, Alex, Andrew, and Arthur. The four women's' names are Anna, Amy, Alicia, and Agnes. One couple ordered Chinese food, the other couple ordered Vegetarian food, the third couple ordered Italian food, and the fourth couple ordered French food. At the end of lunch, they received four bills, one per each family. The amounts on the bills were $45, $59, $69, and $73. Please figure out who is married to whom, what food each couple ate, and the amount each couple paid for the lunch based on the following clues:
   - Ansar paid more than Anna's husband
   - Alicia's husband paid $14 less than Alex
   - Arthur paid more than the man who ordered vegetarian food but less than the man who is married to Amy
   - Anna and her husband ate Italian food
   - Alicia and her husband ate French food
   - Vegetarian food is more expensive than French food

10. **Nicholas, Karin, Elina, and Anna are the best in solving logic problems**. On the last test their marks were 93, 94, 96, and 98 - much higher than the rest of their class. When asked, who got what mark, their answers were:

    Nicholas: My mark is the lowest among four of us
    Karin: My mark is not the lowest among four of us
    Elina: My mark is not the lowest and it is not the highest among four of us
    Anna: My mark is the best among four of us

    The teacher, hearing all these four answers, commented that one of the answers was wrong but three were correct. As well, she added," Anna did better than Elina on this test."

    a. Who among the four students lied?
    b. Can you figure out who got what mark?

Level Three

Fun Home Assignment:

1. **Joseph is driving his car on a highway** to visit his daughter Dana, a student at the university. She lives in a dormitory very close to the highway. They agreed to meet at 12pm at the restaurant on the highway somewhere between Dana's dormitory and the city where Joseph lives. While Dana was riding her bike at her maximum speed along the highway to the restaurant, she received a phone call from her dad when she already made $\frac{4}{7}$ of the distance from her dormitory to the restaurant. Joseph asked Dana if she is bringing with her the latest report card as he needs to give a copy of the card to his insurance agent. Of course, Dana forgot about her promise to give dad her report card and she became very upset about it. She was sure if she would continue riding her bike at the same speed, she would arrive to the restaurant at the same time as her dad. But, if she will turn back and continue riding her bike to her dormitory at her maximum speed and, if her dad will continue driving to Dana's dormitory, then both of them will arrive there at the same time. Then they will have to skip their lunch together at the restaurant. At what speed was Dana riding her bike, if her dad's speed was 91km/h?

2. **How much pure water must be added** to 4.5 liters of 87% acid solution to dilute it to 64% solution? (Round your answer to the nearest hundredth).

3. **A kilogram of raisins cost $6.54** and a kilogram of nuts cost $7.28. How many kilograms of raisins and how many kilograms of of nuts must be taken to prepare 6 kilograms of a mixture that will sell for $7.10 per kilogram? (Round your answer to the nearest hundredth).

4. **Dana invested her savings of $6400**, part in 8% annually, and another part in 9.2% annually. At end of that year Dana received in total $555.2 interest on both investments. How much did Dana invest at 9.2% rate?

5. **How much of juice concentrate**, which tested 78%, should be added to 5.5 litres of filtered water to obtain juice which will be tested 34%?

Level Three

6. **Given**: Tangent AT = 16m;

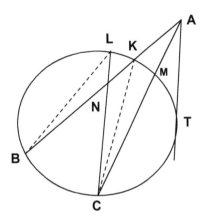

AK = 8m, NK = 4m, AM = 6m, LN =5m.
∠BLC = 38°, ∠LCM = 26°,
Segment CK is bisecting ∠LCM.
Find: a) ∠ BAC
b) ∠ BNC
c) BK
d) LC
e) AC

7. **Two containers hold 620 liters of liquid** each. When drains are open in each container, the first container drains 18 litres per minute and the second container drains 12 litres per minute.
   A. After how much time will the volume of the remaining liquid in one container be half of the volume of the remaining liquid in the other container?
   B. After how much time will the volume of the remaining liquid in one container be $\frac{1}{3}$ of the volume of the remaining liquid in the other container?

8. **Gleb bought tickets to a concert** in another town. While driving from his house for an hour at a speed of 70km/h he realized that he will be late for the show by 15 minutes. Gleb pushed on the gas pedal and increased his speed by 30km/h. He arrived to the concert hall 21 minutes before the concert's start time.
   a. At what time does the concert start if Gleb started driving at 12pm?
   b. What is the distance between Gleb's house and the concert hall?

9. **Can you find such a number X so X% of X is equal to 49?** What are the other whole numbers between 1 and 100 (including 100) that you can find so X% of X will be equal to such a whole number? What is the rule?

10. **Lev and Yaniv are working at a warehouse** from 8am in the morning. Every day they have to load a truck with the daily orders to deliver to customers. Together they can fill the truck in 4 hours and 48 minutes. However, yesterday Yaniv had a dentist appointment in the morning. By the time he showed up for work, Lev already filled up half of the truck and decided to go play golf. Yaniv had to finish loading the second part of the truck by himself. He finished working at 6pm. How much time will it take each of them to fill the truck working alone if it is known that Yaniv is slower at work than Lev?

LESSON 21

Classwork:

1. **A 35% of alcoholic beverage is added** to a 60% of alcoholic beverage to create 50 litres of 42% beverage. How much of 35% alcoholic beverage is taken?

2. **A 74% acid solution needs to be reduced** to 38% solution by adding 22% of acid solution to it. How much of each solution should be taken to obtain 40 litres of 38% solution?

3. **Given two circles with centres in O_1 and O_2**. $\angle A = 24°$. Find values of:
   a. $\angle C = ?$ b. $\angle O_1 KB = ?$

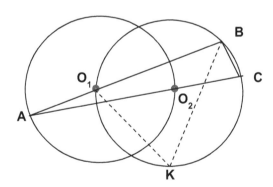

4. **Solve the systems of linear equations** using substitution, comparison, and subtraction methods:

   a. $2x - 5y = -3$
       $3x + 2y = 24$

   b. $x + 4y = 15$
       $7x + 9y = 10$

5. **Two geese and five chickens cost $90**. Four geese cost $24 more than 3 chickens. What is the total price for three geese and 4 chickens?

6. **Selina is now six times older than Nicole was**, when Selina was twice younger than she (Selina) is now. How old are Selina and Nicole now if the sum of their ages 10 years ago was 20?

7. **Two times the tens digit of a two digit number** is one less than five times the unit digit. Three times the tens digit plus twice the unit digit equals 27. What is this two digit number?

8. **12 kg of sugar and 7 kg of barley cost $62.50 last year**. This year there was an increase of 10% on the price of sugar and 8% on the price of barley. The same purchase of 12 kg of sugar and 7 kg of barley will cost $68.4. What were the prices of sugar and barley last year?

Level Three

9. **Four married couples, eight people in all**, met to play Bridge (card game). In the game of Bridge four people sit at the table, two people playing against two people. These eight people are split among two tables so each pair of players/partners consists of one man and one woman. Based on the clues below, please find out who is the spouse of whom and who is playing on each team:
   - The four men's names are Jack, Jerry, John, and Jay; the four women's names are Maria, Emma, Eden, and Elina
   - Maria was on the team with Eden's husband
   - Jack and Elina played against Eden and Emma's husband
   - John's wife's partner in the game was Jerry
   - John's partner was Jack's wife
   - No team of partners consist of a husband and a wife

10. **A motorcyclist started riding from city A to city B at 9 am.** The distance between these cities is 473 kilometers. At a distance of 275 kilometers from city A, motorcyclist stopped at the restaurant on the road and spent 1 hour and 18 minutes to eat and to rest. After that, he continued riding his motorcycle at a speed 20 km/h slower than before and arrived to his destination at 3 pm. What was the motorcyclist's speed before he stopped at the restaurant?

Level Three

Fun Home Assignment:

1. **Solve the systems of equations** with two variables using various methods (substitution, comparison, and subtraction):

   a. $2x - 7y = 10.5$
   $x + 4y = 9$

   $a_1.$ $5x + 4y = 2$
   $4x + 5y = 7$

   b. $3x - 5y = -1$
   $6x + 15y = 8$

   $b_1.$ $0.5x + 2y = -12$
   $2x - 3y = 29$

2. **A container holds 40 litres** of juice containing 40% concentrate. Using special filtration equipment, it is possible to extract from this container juice containing 62% concentrate. How many litres of juice containing 62% concentrate should be extracted from 40 litres of juice containing 40% concentrate to reduce concentration of juice in a container to 30%?

3. **A factory warehouse holds a supply** of 2 kinds of concentrated cleaning solutions, one is 35% concentration and another is 65%. How much of 35% solution should be taken to prepare 50 litres of 42% cleaning solution?

4. **The difference between three times** the first number and two times the second number is 39. When you add two times the first number to four times the second number, the result is 58. What are these two numbers?

5. **Given:** $\angle A = 17°$, $\angle C = 31°$.
   PK = 10 cm, KC = 30 cm, BR = 35 cm.
   AT is tangent to the circle;
   AT = 28 cm, BP = 33 cm.
   Find: $\angle BKC$, BK, KR, and AB

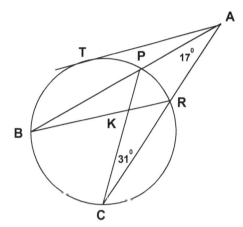

6. **Five students, Sam, Alexia, Daniel, Ben**, and Karen, each majoring in one of the five subjects, algebra, geometry, calculus, trigonometry, and statistics, received an assignment (a project) from their teachers, Brown, Smart, Clark, Rubinshtein, and Bergman. From the clues below, please figure out for each student what subject he/she is majoring and from which teacher this student received an assignment/project:
   • Karen is majoring in algebra
   • Teachers Clark and Rubinshtein are not teaching algebra nor geometry
   • Daniel is majoring in geometry
   • Teacher Smart did not give an assignment to Ben
   • Teacher Clark is not teaching trigonometry nor statistics
   • Teachers Brown and Smart never met Sam, Alexia, and Daniel
   • Sam is majoring in trigonometry

Level Three

7. **A railway track is parallel to a road.** A school bus driver is riding his bus on the road at a speed of 60 km/h. The bus driver noticed a train, 125 meters long, coming in the opposite direction and it passes his bus in 3 seconds. What is the speed of the train?

8. **A given fraction is equivalent to $\frac{3}{5}$.** When 16 is added to the numerator of the fraction and 15 is added to the denominator of the fraction, the new fraction will be equivalent to $\frac{4}{5}$. What is the original fraction?

9. **Given a 2 digit number.** The quotient of the two digit number and the sum of its digits is 4. The difference between the reversed and the original two digit number is 36. Find the original two digit number.

10. **Triangle ABC is isosceles.** Length of sides AB and BC is equal to 12 cm each, and length of the base AC equals 20 cm. Find the length of the radius of the circle that inscribes this triangle (i.e. all three vertices of this triangle must be on the circumference of this circle).

Level Three

LESSON 22

Classwork:

1. Solve systems of equations:

   a. $3X - Y + 5Z = 22$
   $2X + 3Y - 4Z = 0$
   $4X - 2Y + 3Z = 27$

   b. $5X + 2Y - Z = 0$
   $X - 4Y + 3Z = 9.2$
   $4X + 5Y - 2Z = 1.8$

2. **Three students, Elizabeth, Stephanie, and Arthur, did not get great marks** on their algebra test. In addition, they tend to lie about their own mark, each of them exaggerated his/her mark by a factor of three. When asked what the sum of their marks on the algebra test was, Elizabeth said 300, Stephanie said 290, and Arthur said 260. What was the mark on the algebra test for each of them?

3. **The length of a side of a square ABCD** is 10 meters. A circle is inscribed in this square – all sides of the square are tangent lines for the circle. An equilateral hexagon is inscribed into this circle – all vertices of the hexagon are on the circumference of the circle. Find the area of the inscribed hexagon.

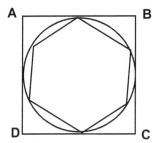

4. **The manager of electronics department** in a department store is preparing report about sales results for the day. Today there was a big discount on TV's, laptops, and on printers. The number of printers sold today was half of the combined number of TV's and laptops sold this day. The sum of the number of laptops and printers sold was two more than the number of TV's sold. The combined number of TV's and printers sold was 10 more than the number of laptops. How many TV's, laptops, and printers were sold today?

5. **A team of three carpenters, John, Jack, and Jay** are representing their company on international competition of carpenters. One of these three men is a manager. The competitors must build a wooden house to the specifications created by the organizers of this completion. John and Jack together can build such house in 10 days. Jack and Jay can build the same house in 15 days. John and Jay together can build the same house in 30 days.
   a. Who of these three men is a manager?
   b. How much time will take each of them to build the house if each of them will work alone?

Level Three

6. **Five women, Natalie, Norma, Emily, Sherry, and Marian** are sitting in the restaurant and bragging about their husbands, Nick, Nir, Mike, Dan, and Max. The husbands' professions are a policeman, an engineer, a University professor, car dealer, and a mechanic. Please find out who is married to whom and what is the profession of each man, based on the clues below:
   - Nick is a mechanic
   - Natalie is nor married to Mike
   - Marian's husband is not a professor nor a car dealer
   - Emily's and Sherry's husbands' names are not Nir, Mike, or Dan
   - Marian's husband's name is not Nir
   - Emily's husband is not a mechanic
   - Max is an engineer
   - Mike is a car dealer

7. **Last year Ma'ayan paid in total $19.70** for 3 kilograms of apples and 4 kilograms of cherries in the supermarket. This year the price for apples increased by 20% and for cherries by 30% due to the bad weather conditions this spring and summer. This year she paid $28.30 in total for 2 kilograms of apples and 5 kilograms of cherries. What are this year's prices for one kilogram of apples and one kilogram of cherries?

8. **Five students, Adel, Ben, Maria, Selina, and Sam** wrote a quiz on logic problem solving. Their marks on this quiz were 75, 80, 85, 90, and 95. When asked who got which mark, they gave these five answers:
   - Ben's mark was 95 and Maria's mark was 90
   - Sam's mark was 90 and Selina's mark was 85
   - Selina's mark was 95 and Adel's mark was 80
   - Sam's mark was 90 and Adel's mark was 80
   - Maria's mark was 85 and Ben's mark was 75

   What was Adel's mark on the quiz if it is known that in each of the answers above one fact is true and one fact is false?

9. **Given a cylindrical container with the circumference** of its base equals 12 meters and its height equals 10 meters. A spiral line, with constant slope, starts at the base and circles exactly twice around this container.
   a. What is the length of the spiral line?
   b. What would the shape of the spiral line will be if you cut the side of the container vertically from the point where the spiral line starts to the point where it ends?

10. **Eric, an experienced carpenter**, was hired to build a huge cottage on a lake. However, after a few days of work, he realised that there is no way he will finish the job during the summer. He called his brother Dave, an owner of a construction company, for help. Dave

sent his best carpenter to help Eric. This also was not enough and Eric asked for more help. So, Dave released more than 2 of his carpenters, one by one, at the same time intervals of 2 days, to help Eric. Each of Dave's carpenters worked with Eric full working days until the cottage was finished. By that time, Eric spent six times more days working on the cottage than the last carpenter sent by Dave. A few days later, sitting in the bar and drinking whiskey with Dave, Eric said that if Dave would have release his carpenters to help Eric from the beginning, it would have taken only 18 full days to build the cottage.
How many days did it take to build the cottage?
How many carpenters did help Eric if David had less than 8 carpenters on his team?

Level Three

Fun Home Assignment:

1. Solve systems of equations:

   a. 3X - 2y + 3Z = 25
      2X - 5Y + 3Z = 26
      2X + 3Y + 5Z = 14

   b. 4X - Y + 7Z = 7
      8X + 3Y - 2Z = 4
      2X + 2Y + 5Z = 8

   c. X - 5Y + Z = 4
      3X + 2Y + 5Z = 1
      2X - Y + 3Z = 2

2. Find a five digit number such as:

   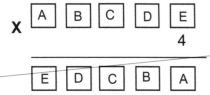

3. Boris, his wife Sharon, and their kids are on a winter vacation in Florida. Yesterday was a very hot day and Boris spent $34.70 on the beach buying for his family two cones of ice cream, three cans of beer, and 4 cans of soda water. The day before yesterday was cooler, so Boris spent only $26.50 on two cones of ice cream, two cans of beer, and three cans of soda water. Today is the last day of their vacation. To make this day memorable, Boris bought five cones of ice cream, two cans of beer, and five cans of soda water. He spent $42.90 today. What is the price of one cone of ice cream? What is the price of one can of beer? What is the price of one can of soda water?

4. A circle is inscribed into equilateral triangle ABC with a side length of 12 meters. An equilateral hexagon is inscribed in this circle. What is the perimeter of the hexagon?

   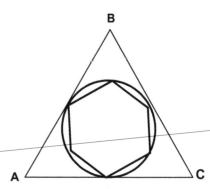

5. The sum of three even numbers is 64. The quotient of one more than the second number and two less than the first number is 0.5. Quotient of the four more than twice the second number and two less than the third number is 2. What are these three numbers?

6. Two teams of road police are patrolling a highway between two cities, A and B. The distance between A and B is 800 kilometers. Both teams usually start driving toward each other, one from city A and another from city B, at 10 a.m. They would meet after driving for 5 hours at a gas

station on the road, would have a snack and coffee together, and then would return back to their cities. One day the team from city A started driving two hours earlier than usual, at 8 a.m., and the team from city B started one hour later, at 11 a.m. The place they met at that day was $118\frac{1}{8}$ kilometers from their usual place of getting together, closer to the city B.
At what speed was each team driving on the highway?

7. **A scientist in the chemistry lab** found a container with an alcohol mixture. When he added 10 litres of 70% mixture of alcohol to this container, the measurement showed 42% percent concentration in the mixture. When he added an additional 10 litres of 30% mixture, the scientist was amazed that this time the mixture measured exactly 40% concentration. How many litres of mixture were in the container originally and what was the original concentration of alcohol in the container?

8. Students in grade 6, Veronica, Alicia, Jane, Sandra, and Alex, are comparing who had collected more sweets on Halloween night. Their last names are Brown, Levy, White, Rubin, and Lee. They each wore different costumes that night: witch, pirate, skeleton, pumpkin, and Spiderman. Each of the students collected different amount of sweets: 60, 82, 85, 95, and 98. Read the clues below and determine each student's full name, her/his costume, and an amount of sweets she/he collected on Halloween night:
   - Rubin wore a pirate costume, but Brown was Spiderman
   - The student wearing a skeleton costume did not collect 98 sweets
   - The number of sweets collected by the student in the pirate costume was less than the number of sweets collected by Brown
   - Lee collected more sweets than White but less than Levy, Rubin, and the student wearing the Spiderman costume
   - Alex collected less sweets than Jane but more than Alicia, Lee, and the student in a pumpkin costume
   - Alicia did not collect 60 sweets
   - Levy collected the most sweets
   - Sandra's last name is not Lee

9. **A circle is inscribed into equilateral triangle ABC** with a side length of 24 meters. An equilateral triangle is inscribed in this circle. What is the area of the shaded part of the inscribed circle?

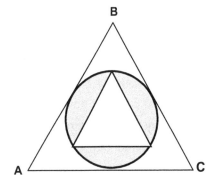

10. **This is an old classic problem** about an interesting argument that happened a long time ago in ancient Greece between a Law School teacher and his student. The student was one of the best in the class but he was so poor that he could not pay his tuition fees. When he came to his teacher to inform that he has to quit because he cannot pay tuition fees, the teacher offered him a deal: The student can continue his studies without paying his fees until he will complete all courses and will become a lawyer. After that, the student will have to repay his fees when the student will win his first case in court. The student accepted the deal and continued his study. However, after receiving a lawyer's certificate, the student married a rich woman and did not practiced the law and never repaid his debt to his teacher. One day the professor came to visit this student and demanded his fees. He said to the student, "I am taking you to court. If you will win the case because the judge will decide that you do not have to pay me, then you will pay me according to our deal! However, if the judge will decide that you have to repay me, then you will pay me according to judge's decision! So, be ready to pay me". "Not so fast!" - replied student. "If the judge will decide that I do not have to pay you, I will not pay you according to the judge's decision! But, if the judge will decide that I do have to pay you, then I will lose my case and, therefore, I will not have to pay you according to the terms of our deal!!"

Who is right? Can you solve this problem using logic?

LESSON 23

Classwork:

1. Solve equations using the formula:
   $$x = \frac{-b \pm \sqrt{b^2 - 4ac}}{2a}$$

   g. $\quad 5X^2 + 9X - 2 = 0$

   h. $\quad X^2 + 3X - 28 = 0$

   i. $\quad 2X^2 - 17X + 30 = 0$

2. A motor boat moved 18 kilometers downstream the river and then another 36 kilometers on the lake. It took 3 hours and 24 minutes for the whole trip. What was the speed of the motor boat on the lake if the rate of the river was 3km/h?

3. Two builders, Frank and Tony, working together can build a cottage in 6 weeks. It takes Frank, working alone, 5 weeks less to build the same cottage than Tony working alone. How long would it take each of them to build the cottage working alone?

4. Evaluate:

   a. $\quad \dfrac{3^2 \cdot 3^6}{3^5} =$

   b. $\quad \dfrac{6^2 - 3^3}{2^2} =$

5. What is the last digit in each of these expressions:

   a. 4^{2014}

   b. 4^{2017}

6. Given two intersecting circles with centres in O_1 and O_2. The length of the radius from O_1 is 6cm and the length of the radius from O_2 is $\sqrt{17}$ cm. Point P is on the straight line going through two points of intersection of these two circles. The distance PO_1 is 10cm. What is the length of PO_2?

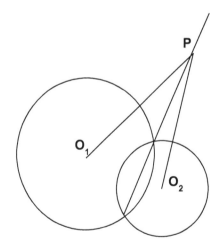

Level Three

7. **Elizabeth made a discovery** playing with numbers. She found an interesting two digit number from which she subtracted the sum of its digits. Then she subtracted from the answer the sum of its digits. She continued doing this until the final answer became 0 (zero). What was the starting number if it took 10 steps to get to zero?

8. **Ted was driving from Toronto to Ottawa.** The time that he was driving in the first 162.5 kilometers was 1.75 hours less than the time it took him to drive the next 225 kilometers. At what speed did Ted drive in each of these two parts of his journey to Ottawa if his speed on the last 225 kilometers stretch was 5km/h less than his speed on the first 162.5 kilometers?

9. **A store manager prepares a quarterly** report indicating earnings of his four employees, Kathy, Mitch, Olga, and Nicole, for the months of April, May, and June. Based on clues below, please find out how much each of the employees earned each month:
   - The April salaries were $500, $650, $720, and $870
   - The May salaries were $400, $650, $700, and $720
   - The June salaries were $700, $720, $870, and $890
   - Mitch's May earnings were less than Nicole's but more than Kathy's
   - Mitch's June earnings were lower than Nicole's
   - Olga earned in May $20 more than the employee who earned $500 in April
   - The employee who earned $650 in May, earned in April $150 more than Kathy
   - The employee who earned $720 in April, earned in June $20 less than Olga
   - Nicole's April earnings were the lowest among employees
   - Each employee's earnings were different in these three months

10. **The length of the rectangle ABCD** is five meters less than three times its width. If the width of this rectangle would be increased by 4 meters and the length would be increased by 6 meters, the area of the new rectangle will be 130 m² more than the area of the original rectangle ABCD. Find the dimensions of the rectangle ABCD.

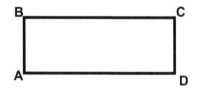

Fun Home Assignment:

1. What is larger:

   a. 2^4 or 4^2 ?

   b. 2^3 or 3^2 ?

   c. 2^5 or 5^2 ?

   d. 3^{3^2} or 3^{2^3} ?

2. Solve the equations:

   a. $6X^2 - 7X - 5 = 0$

   b. $\dfrac{X-3}{2X+7} = \dfrac{X+4}{2X-5}$

   c. $\dfrac{3X+2}{X+4} = \dfrac{2}{2X+1}$

   d. $28X^2 + 23X - 15 = 0$

3. **Two buses left Toronto** to the destination 360 kilometers from the departure point. The first bus travelled with the speed 10 km/h more than the second bus and it arrived 30 minutes earlier to the destination. What was the speed of each bus?

4. **What is the last digit in each of the expressions below:**

   a. 2^{2014}

   b. 3^{2015}

   c. 9^{2016}

   d. $9^{9999^{999999}}$

   e. $4^{5015} - 9^{2022}$

5. Boris and Gabriel are working in the warehouse. They are packaging boxes according to the orders from their clients received yesterday. After working together for four hours, they packaged $\dfrac{2}{3}$ of the big order.

   Boris would package this order alone in five hours less than Gabriel, working alone.
   How many hours would it take each of them to package the order working alone?

6. **Ben's hobby is fishing on lakes and rivers of Muskoka.** He bought a small motor for his row boat which allows his boat to move at a speed 10km/h on lakes. Yesterday Ben ventured to go 42 kilometers up the river and back. Ben noticed that it took him 4 hour more to go upstream than downstream. What was the rate (speed) of the river?

7. **Five students, Rick, Bob, Michelle, Carol, and Vicki** are sitting nervously around a table. Their last names are Green, Johns, Cohen, Levy, and Chen. They were invited to discuss with the teacher's committee the achievement

awards they may receive at the end of the school year. This year their school will award special prizes for achievements in Math, Science, History, Music, and Art. While waiting to be called in, the students put on the table a book, a notebook, a cellphone, a puzzle, and a binder.

Based on the clues below, please find out each student's full name, item he/she put on the table, and for what subject each of them is expecting an award:

- Bob does not expect an award in Science nor in Music
- Rick expects an award in History
- Green's first name is not Michelle nor Carol
- Carol knows that she is going to get the award in Music
- Either Michelle or Carol put notebook on the table
- Chen is the best student in Art and she brought a notebook
- Green talked on the phone and is expecting to get an award in Science
- Not Bob and not Carol put a book on the table
- Johns is busy with solving his puzzle
- Bob is not Levy
- Cohen expects an award in Music

8. **The tenth digit in a two digit number** is three times the unit digit. The product of this two digit number and the sum of its digits is 1116. What is the sum of the original two digit number and the reversed number?

9. **David is driving his car on a road** at a speed of 40km/h. A train, 144 meters long, is going in the direction towards David on the railway which is parallel to the road where Dave is driving his car. David noticed that it took exactly 6 seconds for the train to pass his car completely. What is the speed of the train?

10. **This is a teaser my teacher challenged me** more than 50 years ago in our after-school math program:
A mountain climber managed to scale up a 100 meters high cliff in a shape of a tower. On top this cliff grows a tree. As well, another tree is growing from the side of the cliff at the height of 50 meters from the ground. Due to the heavy wind and rain, the mountain climber lost all his equipment except a knife and a 75 meters rope. The cliff is in a deserted area and there are no people around to help the climber to get down. After some time of thinking, the climber found a simple solution for getting down unharmed. How did he get down?

Level Three

LESSON 24

Classwork:

1. **Steve is observing a long train** passing a tree in 20 seconds. As train is approaching a bridge 380 meters long, Steve noticed that it takes 60 seconds for the train to pass completely through the bridge. At what speed the train is moving?

2. **Exponents and Radicals:**
   a. How many different factors are in the exponential expression: $7^5 \cdot 49^5$?

   b. Which expression has a higher value: $\sqrt{2012} + \sqrt{2014}$ or $2\sqrt{2013}$?

   c. **Solve system of equations:**
   $X^2 + Y^2 = 13$
   $XY = 6$

3. **Given circle with centre in O.** Point A is outside of the circle. Line AC passes through the centre O. Segment AT is tangent to the circle in point T. length of the segment AB is 12 meters and length of tangent AT is 24 meters. Find the perimeter of the equilateral hexagon inscribed in the circle.

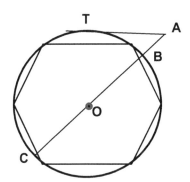

4. **A lab technician is asked to determine** concentration of acid solutions in two large barrels. The technician noticed that labels with needed information on these containers are missing. She decided to use two empty jars 20 liters and 30 liters to find out the levels of concentration of acids in the barrels. When she mixed up 20 liters from first barrel with 30 liters from the second barrel, the resulted mixture had 51.2% of acids. But when she mixed up 30 liters from first barrel with 20 liters from the second barrel, the resulted mixture had 55.8% of acids. From that, the lab technician was able to calculate the concentration level of acids in each barrel. What these numbers were?

5. **The unit digit of two digit number** is 3 more than the tens digit. Sum of squares of these digits is 31 more than this two digit number. Find the original two digit number.

6. **The mother of the school principal's daughter,** during a phone conversation with the daughter of the school principal's mom, asked her to congratulate her husband with their anniversary. The woman passed the message to her husband; however the school principal never received this "Congratulation" message. How could this be?

Level Three

7. **Mrs. Berger is a very busy mom.** Besides having five school-age children, she is a Vice President in a large manufacturing company. Today she was late for work so she needed to pack her children's' lunch boxes very fast and she put these lunch boxes in their backpacks. At school, her children realized that their mom mixed up everything: each lunch box had wrong sandwiches and wrong salads. As well, sandwiches and salads in each lunch box belong to different children. Now the children want determine in which lunch boxes are their sandwiches and salads? Please help Mrs. Berger's children, Jake, Jordan, Judy, Jane, and Joseph to solve this puzzle if:

   - Jane's salad is in Jordan's lunch box
   - Jake's salad and Judy's sandwich are in the same lunch box
   - Jordan's sandwich and Joseph's salad are in different lunch boxes
   - Jane's lunch box has Joseph's sandwich but not Judy's salad

8. **The bus driver is assigned to the route** that is 20 kilometers long. On his last ride after mid-night, the driver increased his speed by 5km/h and he brought his bus 12 minutes earlier than usually to the last stop. What is the bus's regular average speed on this route and how much time it take the bus to cover the distance of 20km?

9. **It takes two teams of workers 4 days** to complete $\frac{2}{3}$ of an order from a very important client. How many days it will take each team, working separately, to complete the order if the more experienced first team can complete the order in 5 days less than the second team?

10. **George lives on town A but works in town B.** John lives in town B but works in town A. Usually they walk to their workplaces by the road connecting towns A and B. Yesterday, George left his town to work at 6am and John left his town to work at 7am. They met on the road at 8am but both continued walking without stopping. George arrived to his workplace 28 minutes later that John arrived to his workplace. Each of them walked at a constant speed and did not stop walking until arrived to work. **How much time each of them spent walking that morning?**

Level Three

Fun Home Assignment:

1. Solve systems of equations:

   a. $(X - 1)(Y - 1) = 3$
      $(X + 2)(Y + 2) = 24$

   b. $X^2 + 2Y^2 = 17$
      $X^2 - 2XY = -3$

2. A manager at a factory received a daily order to assemble 175 bicycles. The manager assigned each of his employees to this project; each employee had to complete the same number of bicycles. However, due to emergency, two of the employees had to be transferred to another project. As result of this, the project was completed by the rest of the team. However, each member of the team had to assemble 10 more bicycles than their original assignment was. How many employees are reporting to this manager?

3. Draw the shapes below in one continuous line without removing pencil from the paper and without going twice over the same line (you can cross lines!):

   a.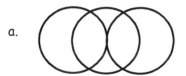

   b.

   c.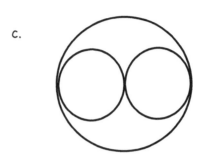

4. X is any prime number greater than three (3). Prove that expression $X^2 - 1$ is divisible by 24.

5. Dana is a chemistry lab technician. She is working with 2 containers with acids; acid concentration in these containers is different. When Dana mixed 25 litres of acids from the first container with 30 litres of acids from the second container, the resulting mixture was 10% concentration. When Dana mixed 50 litres from the first container with 20 litres from the second container, the resulting mixture was $12\frac{6}{7}\%$ concentration.

   What is the acid concentration in each container?

6. **Do you remember witches that captured 10 students (in the Level 2 Book Two)?** Two of these old ladies decided to improve their logical thinking and they applied to Brain Power Logic Development program. To get registered, they had to pass a simple test. **The first witch was given the following problem:**

   a. Assume you are captured by a police force and you are accused of mistreating poor students. You put in front of a judge who is saying that you will be punished harshly for your horrible past. The judge gives you a choice to spend the rest of your life in the prison with very aggressive cannibals or be locked in a large slow boiling pot. You can have a choice of punishment by saying a simple phrase/sentence. If your phrase will be a true statement, then you will be spending your life with cannibals. However, if your phrase will be a false statement, you will be boiling in the pot to the end of your life. You will be accepted to Brain Power program if you will construct a phrase that will get you out of your predicament!!

   The second witch was given the following problem:

   b. Let say you have captured one of our students. The parents of this student are begging you to return their child. Let say, you promised them sincerely that you will return the student if parents will guess correctly if you will or will not return the child. What will you do if parents will guess that you will not return the child?

   Which of these witches have more chances to get accepted to Brain Power?

   Could both witches be accepted? If your answer is YES, then what answers the witches must provide?

7. **Students in Brain Power programs decided to spend summer time helping building small cottages** in the forest area near the ocean. Students from two classes could build a cottage in 8 days working all together. However, if only $\frac{2}{3}$ of students of the one class would work together with $\frac{4}{5}$ of the students of the other class it would take $11\frac{1}{4}$ days to build the same cottage.
   In how many days each class alone can build this cottage?

8. **Two classes were paid for building** cottages. Since the cottages were the same, each class received the same amount of money per cottage. There were two students less in the first class than in the second. Therefore each student in the first class received $1 more than each student in the second class. Both classes together received $26 more dollars than the total number of students in both classes.
   How many students are in each class?

Level Three

(Do not be surprised by the "low" pay. The students were paid in golden coins!!)

9. The radius of one circle is 10 meters and the radius of another circle is 5 meters. These circles touch each other externally in the point M. A tangent line touches these circles in points A and B (see the picture below).

a) Prove that the tangent line common to these circles, drawn through the point M, splits the segment AB into two equal segments.

b) Find distance from point M to the segment AB.

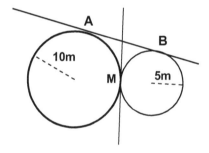

10. Find two three digit numbers if:
a. Both numbers contains the same digits, but the hundreds digit in the first number is the same as the unit digit in the second number, and the unit digit in the first number is the same as the hundreds digit in the second number. The tens digits in these numbers are the same.

b. The sum of these two three digit numbers is 1147.

c. The sum of the digits in each of these numbers is 14

d. The sum of the squares of all digits in each of these three digit numbers is 78.

Level Three

Level Three

LESSON 25

Classwork:

1. **A musical band is setting the stage** for a concert. Two feet in front of the stage they stuck a pole vertically in the ground. The top of the pole is 1 feet higher than the stage. One clumsy worker touched the pole and it fell toward the stage. Now, the top of the pole is touching the top edge of the stage. What is the height of the stage?

2. **The cost of printing books is a fixed** amount for print setting, plus a fixed rate per book. Jane ordered 9 books and she paid $81 in total. Arthur ordered 17 books and he paid $129. How much did Brett pay if he ordered 23 books?

3. **Five students, Amy, Anne, Alon, Aric, and Ally, bought tickets** to a classic music concerts. Each student bought only one ticket and each concert is on different date in October. Please figure out who is going to what concert and on what date:

   • Ally is going to the concert that is a day before the concert to which Alon bought his ticket
   • Beethoven's music concert is a day before Mozart's music concert
   • Aric is going to the concert one day before Beethoven's music concert
   • Amy is going to the concert that is two days before Bach's music concert
   • Bach's music concert is one day earlier than Mendelsohn' concert
   • Anne does not like Mozart's music

	Concert					October dates				
	Mozart	Bach	Beethoven	Chopin	Mendelsohn	1	2	3	4	5
Amy										
Anne										
Alon										
Aric										
Ally										
1					
2					
3					
4					
5					

4. **Three water pipes opened together can fill in the large water tank in 6 hours.** Pipe B alone can fill the water tank in $\frac{3}{4}$ of the time it takes pipe A alone to fill this water tank. It takes pipe C alone to fill the water tank in 10 more hours than pipe B opened alone. How much time it take to fill the tank if, first, pipe A alone will be opened for one hour, then pipe B will be opened in addition to pipe A for another hour, and then pipe C

will be opened in addition to both pipes A and B, until the water tank will filled in?

5. **Two circles with centres in points O_1 and O_2 intersect in points A and B.** The length of AB is 6 meters. The radius of the circle with centre in O_1 is 7 meters, and the radius of the circle in centre O_2 is 12 meters. Calculate the area of quadrilateral O_1AO_2B.

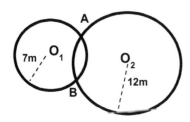

6. **The unit digit in the original three digits number is three times larger than the tens digit.** Sum of the digits in the original three digit number is 13. In the reversed number, the digits changed positions in the following way: the tens digit became hundreds digit, the unit digit became tens digit, and the hundreds digit became unit digit. Twice the reversed number is 4 (four) more than the original number. What is the sum of the original and the reversed numbers?

7. **At the beginning of the first lesson**, the classroom teacher Mr. Gallagher is marking students' attendance. He noticed today that the number of students absent is $\frac{1}{6}$ of the number of students that arrived on time. Few minutes later, one student had to leave the class. After that, the number of absent students became $\frac{1}{5}$ of the number of students present in the class. How many students are in Mr. Gallagher's class in total?

8. **Square ABCD and a circle have a common centre in point O.** The radius of the circle is 1 meter. Area of the square ABCD and area of the circle are equal. What is the area of the rectangle AMTD?

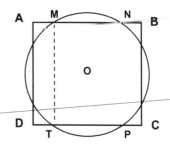

9. **A boat can travel in still water 6 times faster than the rate of river.** City A and city B are located at the river, 140 kilometers apart. The boat travelled from city A to city B and back in 12 hours. Find the speed of the boat in still water.

10. **A store owner is preparing her merchandise for Halloween.** She has soft candies, $3.70/kg, and wrapped chocolate squares, $4.10/kg. The store owner received an order to prepare 10 bags, 1 kilogram each, of candies and chocolates mixed together, so the price per one bag should be $4.00. How many kilograms of candies and how many kilograms of chocolates must be used for this order?

Level Three

Fun Home Assignment:

1. **Stephanie and her friend Elizabeth** went to a school supplies store to buy stuff for the school. Stephanie returned 2 binders she bought yesterday as these were wrong colour. She received money for the returned binders that was exactly enough to buy three pens and one geometry set. Elizabeth bought three binders, two geometry sets, and one pen. She spent exactly $25 for all her items. What is the price for one binder, for one pen, and for one geometry set?

2. **Nick forgot the page number on which was his homework in Math.** He knew it had to be either page 39 or 40, but he could not remember which one it was. Nick decided to give a phone call to his friends to find out the page number for his homework. He has three best friends, Oleg, Stefanie, and Maria, but only one of them is a truth-teller and other two of them are liars. Nick is not sure either, who of his friends is a liar and who is a truth-teller. So, Nick call each of them and got these answers:

   Oleg: "The homework is on page 39."
   Stefanie: "Oleg is lying. The homework is on page 40."
   Maria: "Stefanie is correct. The homework is on page 40."

   On what page is Nick's homework assignment?

3. **This problem I solved at my high school's after-school Math program more than 50 years ago:**

   A man with a hat on his head swims up the river. There are two bridges in front of the swimmer, 2km apart. The man passed the first bridge and is approaching the second bridge. When the man was exactly under the second bridge, he lost his hat. Twenty minutes later the man noticed that he missed his hat. Immediately he turned back and swam to catch up his hat. He caught his hat exactly under the first bridge. Assuming that the swimmer used the same effort while swimming, what is the speed of the river current?

4. **AB is a diameter of the circle with the centre in point O. TP is** a perpendicular to AB (TP $\perp$ AB). Point T is on the circle and point P is on the diameter AB.
   Prove that TP = $\sqrt{AP \times PB}$

   Note: There are more than one way to prove this

   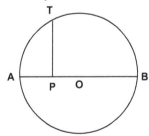

5. **There are 50 litres of a pure acid** (100%) in a container. Some volume of acid was taken out of the container and replaced with pure water. Then the same

Level Three

volume of mixture was taken out as before, and it was replaced with pure water. As result of these two steps, the container has now 50 litres of 64% mixture of acid and water.
What volume of acid was taken out of the container in the first step?

6. **Tom and his 2 little brothers**, Jim and Jack, helped elderly people on their street to clean backyards from dry leaf. At the end of the day, the boys counted money each received for their work. Jim got $12 less than Jack. Tom got twice the amount Jim and Jack received together. How much money did each of the boys receive that day if product of Jim's and Jack's amounts was $104 more than Tom's amount?

7. **Five people met at a vacation resort** in Jamaica, Stephanie, Arthur, Alex, Elizabeth, and Sam. They had a great time dining together and sharing their life experiences. In these discussions, they figured out that they are from five different Canadian cities, Ottawa, Toronto, Winnipeg, Montreal, and Halifax. Their professions are accountant, programmer, engineer, teacher, and physician; each of them has different profession. Please figure out profession of each person and city he/she is from if the following is known:
   - Sam is not a teacher or an accountant. He is from Halifax.
   - Arthur never wanted to be a teacher
   - Programmer is from Montreal
   - Stephanie is not a teacher or a programmer. She is not from Toronto.
   - Elizabeth is a physician
   - A person from Ottawa is a man.

8. AC is a diameter of the circle below. Point B is on the circle. Given that BK $\perp$ AC, MK $\perp$ AB, and NK $\perp$ BC. As well, AK = 4cm and KC = 9cm.
Find the length of MK and of NK.

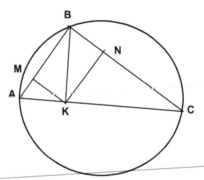

9. **On their first date, Linda asked Robert about his age.** Robert said that her question is not very polite, however he can give Linda a simple problem and she will be able to determine his age when she will find the correct answer to the problem. The problem formulated by Robert was: "The sum of your age and mine is 63. I am now twice older than you were when I was as old as you are now."
How old is Robert now?

10. **Among 13 coins one is faked and the rest are good.** It is not known if the faked coin is lighter or heavier than the regular coins. In addition, there is 14th coin which is a good coin. You have an old fashion scale without weights. Can you find the faked coin using the scale only

Level Three

three times? As well, can you determine if the faked coin is lighter or heavier than the regular coins?

11. **Kathy is at her grandparent cottage on the lake this summer.** Kathy found there two empty containers, 17 litres and 5 litres volume. She needs exactly 13 litres of water. How can she solve the problem if there are no measurement marks on these two containers?

About the Author

Reuven received his master's degree (math and education) from University of Odessa, former USSR. Reuven worked on postgraduate studies and research in mathematics at the Hebrew University in Jerusalem, Israel, and in Paris, France. For over thirty years, Reuven has enjoyed teaching mathematics in four countries around the world, as well as in Canada, and has worked as vice principal in Paris, France. As founder of Brain Power Enrichment Programs in metro Toronto, Reuven continues to lead the school as principal, curriculum developer, and head instructor.

Karine received her undergraduate degrees (BSc and BEd) and completed a master in education at the Ontario Institute for Studies in Education (OISE) at the University of Toronto. Karine's master work focused on teacher development and the field of social justice in the education of immigrant students. Karine completed her doctoral studies in education at York University. In addition to teaching mathematics and science in various schools across Toronto, Karine has been contributing her talents as a specialized Level 1 and Level 2 course instructor at Brain Power Enrichment Programs for nearly ten years. Karine dynamically works in the capacities of administrator, curriculum developer, and instructor of the Brain Power Language Arts Enrichment courses.

Lightning Source UK Ltd.
Milton Keynes UK
UKHW02f2243150718
325725UK00001B/5/P